PRO ... ICY

... iation Guide

The Design Council

Published in the United Kingdom by
The Design Council
28 Haymarket
London SW1Y 4SU
071-839 8000

Printed and bound in the United Kingdom by Staples, St Albans
Edited for the Design Council by Victoria Felton
Designed by Nicole Griffin
Assistant designer Carol Briggs

Illustrations contributed by Sharp Practice
Waterside
44-48 Wharf Road
London N1 7UX
071-490 1712

Every effort has been made to check the factual accuracy of the information given in this book, at the time of publication, but it should not be regarded as a substitute for seeking appropriate professional advice as and when necessary. The Design Business Association and/or the Design Council will not be responsible for any claim arising from action on the basis of information contained in this book. The opinions expressed are those of the individual authors, and do not necessarily represent the views of the Design Business Association and/or the Design Council.

British Library Cataloguing in Publication Data
Professional Practice in Design
Consultancy: Design Business Association Guide
I. Lydiate, Liz
745.4

ISBN 0 85072 304 3

Acknowledgements

The Design Business Association and the Design Council gratefully acknowledge permission to reproduce copyright material from the following individuals and publishers: Peter Drucker and Heinemann for the quote on page 138; Bob Dylan and Sony Music for the lyrics on pages 20 and 148; P Kotler and Prentice Hall for the quote on page 96; Thomas Peters, Robert Waterman and HarperCollins for the quote on page 182; Laura Mazur and Economist Intelligence Unit for the excerpt on page 162.

CONTENTS

FOREWORD

'People are our most important assets.' How many times have you heard that said in the design industry? But do we really mean it and believe it? If we do, what action are we taking to protect, nurture and increase the value of those assets?

Until recently, the design industry track record in training was almost non-existent. Unlike most other industrial sectors we have only just begun to construct training courses designed specifically for people in our sector. It isn't that we don't care. In fact, I know that most design businesses are genuinely caring and concerned about helping their staff to develop. Wanting to do it, though, is one thing; putting the ideal into practice is another because of the negative factors of time, money and fear.

Design businesses spend most of their time and energy worrying about their clients' problems rather than their own. Time set aside for developing our own people can easily and instantly be replaced by a client's needs or demands. Money allocated for the same purpose can all too quickly be used for something seen to be more urgent at the time. And the nagging worry is always there that time and money spent on your member of staff today might contribute to your competitor's success tomorrow. I know how hard it is to overcome these negative factors. But we have got to overcome them if our industry is to move forward and become better and stronger.

I believe we have to do two things. First, we must think long term. Listen to what Akio Morita says: 'In the long run, your business and its future are in the hands of the people you hire. To put it a bit more dramatically, the fate of your business is actually in the hands of the youngest recruit on the staff.' I believe that if we take the long-term view and empower our people by putting the weapon of learning in their hands we will make them stronger and we will make our businesses stronger.

Second, we must put aside the dry and boring association too often attached to the notions of 'professional practice' and 'training'. Let's think and talk instead about the excitement of exploration, discovery, and the realization of potential.

But people rarely become stronger and better by chance. They have to have the will and they have to be given the opportunity. We can give them that opportunity by recognizing that training is fundamental to the future of our industry.

John Sorrell
Chairman
Design Business Association
March 1992

INTRODUCTION

This book is intended to function in at least three different ways. First we hope that it will become a standard reference book for use in design consultancies, and a source of guidance on particular issues and points as they come up in day-to-day work.

Second, it is the textbook for the Design Business Association's Professional Practice Courses, which were established in 1989 in response to members' training needs, consistent with the Association's mission statement which says:

> 'The mission of the Design Business Association is to help member consultancies achieve the highest professional standards and in doing so help their clients get the best possible results from using design.'

The publication of this book extends the effectiveness of the courses by providing a permanent source of reference, but also adds greatly to the material which can be covered. As it is, course delegates embark on the design equivalent of an SAS yomp across Dartmoor. In future, they will also have an ongoing reference manual to take away with them and digest in more leisurely circumstances.

Finally, the book has been put together in a way which is also suited to private study. Individual chapters and their associated reading form self-contained study projects, and can be linked together to cover wider areas of activity.

Another central decision has been the multiple authorship of this book. The design business is diverse by its nature, and there is no single 'right way' to proceed. The important thing is to find the best way in any set of circumstances, and this is a theme which runs through the work of all the contributors. Collectively, the chapters represent over 200 years of working experience in the forefront of design practice.

As editor, I should like to thank all the contributors for work, support and generosity which went many times beyond the call of duty. For all of us, the fact that the book is now in your hands is a source of great satisfaction and the best reward.

In writing for designers and about design practice, we were mindful of Alice's comments on the uselessness of a book without pictures, and made an early decision to convey our messages visually as well as verbally. We are immensely grateful to Sarah Culshaw and the illustrators from Sharp Practice whose generosity has enabled us to turn that concept into reality to an extent and a level we hardly dared hope for.

Finally, I would like to hand the overall credit to Vicky Sargent, the DBA's Chief Executive without whom this book would never have happened. Apart from raising the funding to make the book possible, Vicky has played a key role in helping me choose our contributors, make some difficult editing decisions and survive the slog of proof reading.

This project is yet another example of the DBA's determination to meet the needs of its members and to work in support of the design industry as a whole.

Details of the DBA Professional Practice Courses programme and other information about Design Business Association activities may be obtained from the DBA at 29 Bedford Square, London WC1B 3EG.

Liz Lydiate
March 1992

CONTRIBUTORS

Jeremy Myerson

Jeremy Myerson is a leading design journalist, editor and author. He has been widely published in newspapers and magazines in the UK and overseas, including *The Times, Financial Times, The Guardian* and *Daily Telegraph*, and his books include the *Conran Design Guides* and *Gordon Russell, Designer of Furniture.* He was the founding editor of *Design Week*, which he established in 1986 as the world's first weekly news magazine for designers and their clients, and he is also a former editor of *Creative Review* and senior editor of *Design* and *The Stage* and *Television Today.* In 1989, he set up his own editorial, research and information consultancy Design Intermedia. He is a Fellow of the Royal Society of Arts, and he has contributed to programmes about design on BBC Radio 4 and Channel 4 in the UK and CBS TV in the USA.

Eric Schneider

A former architect, with a first degree in economics and an MBA from London Business School, Eric Schneider now combines running his creative management consultancy, Schneider & Partners, with design management teaching. His academic roles have included a visiting fellowship in the Management of Design at Kingston Business School, teaching at the University of Industrial Arts in Helsinki, as well as a variety of external advisory activities. He has written and edited various works on the management of design, is a regular conference speaker, and was a judge for the1990 DBA Design Effectiveness Awards. Schneider & Partners specializes in providing business planning and marketing to the creative sector, where clients include many design consultancies and architects, and in assisting organizations in the effective planning, organization and management of their products, communications, environments and identities.

Henry Lydiate

Henry Lydiate is a barrister with a special interest in the law in relation to the visual arts. Following a major research assignment in this field, on behalf of the Calouste Gulbenkian Foundation, he established Artlaw Services, a charity which offered specialized legal advice and help on art and design matters from 1976-84. He is the author of *The Visual Artist's Copyright Handbook* and contributes articles to national newspapers, legal journals and art magazines. He is a regular lecturer on art law subjects, fitting this in with his full-time practice in the courts.

Joe Tibbetts

Joe Tibbetts is a founding partner in International Design Marketing, a strategic consultancy which advises design companies at board and senior management level. He writes regularly on design and architecture, and is the author of *Design into Europe.* The text of his address, 'Strategic imperatives for European design companies' given at the Paris Conference in 1990 has been published in Spain, France and Germany. He has worked closely with the DBA to devise and produce its programme 'Developing Business in Europe'.

Bob Willott

Bob Willott's involvement with entrepreneurs and people businesses spans thirty years. His early years in West End practice were oriented to the entertainment industry. Then he joined Haymarket Publishing Ltd to launch *Accountancy Age* followed by a period as a publisher and director. For ten years prior to founding Willott Kingston Smith, he was a partner in Spicer & Oppenheim (latterly Touche Ross & Co), fulfilling technical and marketing roles as well as advising a growing number of clients in the media and marketing sectors. During that period he also produced seven editions of an annual survey of advertising agency profitability, initiated the Financial Communication Awards and played a prominent role in founding the Design Business Association.

Shan Preddy

The principal of The Preddy Consultancy, Shan Preddy specializes in the design business. She is both a marketing and in-house training consultant to design companies, and an independent design manager for manufacturing and service companies, advising them on policy, strategy, and implementation. In conjunction with the DBA, she holds regular open training workshops. Shan has over seventeen years' experience in communications, including public relations and advertising, and was previously managing director of the Tyrell Company, a graphic design consultancy. Shan is a member of the Marketing Society and the Institute of Directors. The Preddy Consultancy is a member of the DBA.

Liz Lydiate

Liz Lydiate studied Fine Art at Newcastle University and St Martin's School of Art, and holds the certificate of the Communications, Advertising and Marketing Foundation with distinction. From a pioneering starting point in the mid-1970s, she has been influential in the introduction of professional practice teaching in art and design colleges nationally. Between 1981-87 she was head of public relations for the Crafts Council, before becoming director of special projects for the Michael Peters Group. She acts as course director for the DBA Professional Practice Courses, and is currently carrying out research for the Chartered Society of Designers as well as consultancy work for design companies.

David Jebb

David Jebb is the senior associate of David Jebb and Associates, a management consultancy he established in 1988, and which works specifically within the design sector. An industrial engineer by profession, David became a management consultant with the PE Consulting Group specializing in profit improvement, the structure of organizations, and management information systems. Later he went into the fashion and quality end of the clothing and knitwear industries where he held senior management positions with Levi Strauss, Sabre, and Dawson International Textiles.

David Rivett

After graduating from the London School of Economics David Rivett joined an international accounting practice and then spent six years running his own business. He gained an MBA from the London Business School where an interest in the management and marketing of design led to his becoming joint managing director of Wolff Olins, and subsequently group development director for Fitch RS Plc. In his five and a half years at Fitch he was responsible for its US expansion, its move into corporate identity, and the opening of European offices. He currently works as a freelance consultant.

Sue Young

Sue Young spent eight years in sales and marketing with ICI before setting up her own consultancy practice, which specializes in marketing and human resource issues. As well as working with large, blue chip clients, she also works with smaller businesses, and in the last four years has been recruiting for the design industry and consulting on management issues in design. She is managing director of Stratego, the recruitment consultancy which she co-founded in 1989, and is an associate consultant of International Design Marketing.

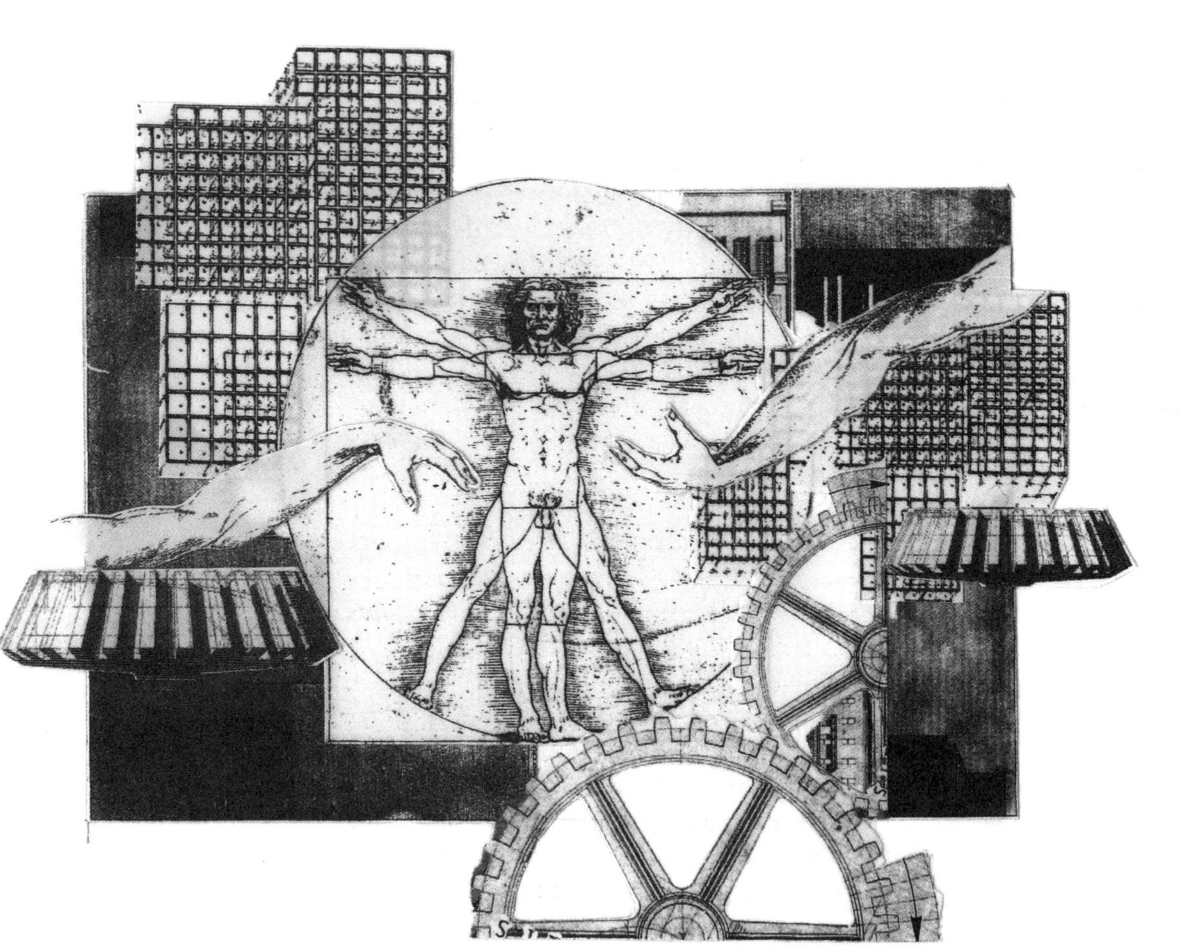

David Loftus/Sharp Practice

'Life without industry is guilt, and industry without art is brutality.'

John Ruskin

1 HISTORY, STRUCTURE AND GROWTH

Jeremy Myerson

Contrary to popular myth among the young designers who grew up fast during the boom years of the 1980s, the British design industry wasn't created overnight when, in the words of Stephen Bayley, 'the *Sunday Times* discovered the Milan Furniture Fair in 1981'. Design has been around in Britain for a very long time, even if our traditions have always leaned more to the literary than the visual. And designing for the needs of industry and commerce has a long history too, even if design for business has often been confused with design as a cultural phenomenon.

How did Britain, with its strong streak of puritanism and deep distrust of the aesthetic, become home to the largest and arguably the most sophisticated community of design consultants in the world? A glance at the history of British decorative art or architecture will reveal that commerce has always accompanied culture.

Creative flair and business acumen

The great patrons of the past decorated interiors, built monuments and commissioned objects to reflect their material status. In their service were designers whose combination of creative flair and business acumen must give many contemporary design practitioners a distinct feeling of *deja vu.*

The classicist architect and interior designer Robert Adam (1728-1792) is one such example. His rich, ornate and unified interiors from the late-eighteenth century survive to this day at Osterley Park and Syon Park in west London. In the 1770s Adam's business employed 2,000 people and, according to records of the time, his senior crafts people (cabinet makers, sculptors and decorative artists) did rather well financially out of his patronage by the rich and influential. A business of that size puts many of today's large communications supergroups in perspective.

The first modern industrial designer

A century later in Victorian times, the man described by many critics as the first modern independent industrial designer emerged. Christopher Dresser (1834-1904) was able to boast in the 1870s that 'there is not a branch of manufacture I do not design patterns for'. A Glasgow-born botanist, he had turned to design after an unsuccessful application for the chair of botany at University College, London.

From 1862 onwards, Dresser gradually built up a large and thriving freelance practice, designing glass, ceramics, tableware and wallcoverings for popular use. Many of his most successful domestic designs were based on metaphors of botany, exploring the intricate constructions of plant life. Tile maker Doulton's, a supplier to department store Liberty's of London, and Sheffield silver plate company Elkington's were among Dresser's clients. According to critic Catherine McDermott: 'Dresser was a significant figure because he is an early example of a designer who was commissioned to create new designs specifically to boost sales for a company.'

Christopher Dresser studied at the Government School of Design from 1847 to 1854. This brought him into contact with Henry Cole and other design reformers of the mid-Victorian age, who were concerned to introduce artists into industry. Working with Prince Albert, Cole, a civil servant, was an influential figure in establishing an institutional framework for design within British public life. In 1848 Cole established a *Journal of Design*, which in a sense puts the magazine publishing launches of the 1980s, such as *Design Week* and *Blueprint,* into historical context. Cole also exhibited his own designs in the 1851 *Great Exhibition* in Hyde Park, a spectacular showcase for the new ideas and inventions of the Industrial Revolution, innovation later to be harnessed by the technological agenda of the twentieth century.

The *Great Exhibition* coincided with Dresser's training as a designer and many of Cole's ideas can be seen in his work. Dresser can also be seen as a bridge between the High Victorians and the Modernists as many of his later metalware designs, such as electro-plated teapots and soup tureens, showed a utilitarian interest in functionalism which prefigured the central design ideology of the Bauhaus by 40 years.

The commercial myth of the Bauhaus

For many designers, the roots of the British design business are to be found in the Bauhaus, the famous German art school. This was founded by the German architect and design theorist Walter Gropius (1883-1969) at Weimar in 1919, and closed down by the Nazis at Dessau just 14 years later in 1933. Yet it has enjoyed an influence on twentieth-century design out of all proportion to its longevity as an institution.

During those 14 years the Bauhaus had only 1,250 students and 35 staff, yet it emerged from the maelstrom of new ideas in fine art in the early years of this century to become the fountainhead of the Modern Movement in design and architecture.

The Bauhaus can claim an impact on post-war British design practice and education in terms of the dissemination of a core methodology and set of beliefs. But the idea that Gropius, Marcel Breuer and other Bauhaus ideologists are the forerunners of the modern international design business does not ring true if related to the ideals on which the Bauhaus was based. Gropius brought painters such as Klee and Kandinsky, sculptors, crafts people, poets and designers together to cross-fertilize ideas in the Bauhaus workshops, to share in the vision of 'a happy working community such as had existed in the mason's lodges of the Middle Ages'.

This ideal forms a surprising link between the Bauhaus, with its glorification of the machine aesthetic, and the Arts and Crafts Movement of William Morris (1834-1896), a committed British socialist, designer and utopian thinker who despised the exploitation of industrial capitalism and saw only misery in machine production.

Like his contemporary Charles Rennie Mackintosh (1868-1928), William Morris produced outstanding designs which are still in production today, ironically earning profits for companies around the world. The Bauhaus was also responsible for training such famous industrial designers as Marianne Brandt and Wilhelm Wagenfeld, whose products, first created under tutor Laszlo Moholy-Nagy in the art school's metal workshops, are now auction-room classics and frequently reproduced.

Although Wägenfeld went on to run his own Stüttgart studio, working for many clients in the post-war years, neither the luminaries of the Bauhaus nor the grand names of British design history can lay claim to be the godfathers of the international design business. That distinction goes to a group of American designers working in New York in the 1920s and 1930s, whose achievements laid the ground for the Anglo-American design groups which emerged more than 50 years later.

America's pioneer design consultants

Best known among the pioneering American design consultants of the period were Raymond Loewy, Henry Dreyfuss, Walter Dorwin Teague, and Norman Bel Geddes. They were very different in background and outlook from the gifted Bauhaus intellectuals who brought European Modernism across the Atlantic in the 1930s in fleeing from Nazi persecution. The Americans' approach was unashamedly commercial, in sharp contrast to the more purist concerns of Walter Gropius, Marcel Breuer and their disciples.

The Americans had come to industrial design from diverse backgrounds in illustration, advertising, stage design, fashion design and store display. They believed that design was a business tool which should be exploited by managers, not an instrument for social and moral change, as Gropius had once argued. While the Bauhaus had explored unadorned, utlitarian, austere forms, the American consultants weren't afraid to use streamlining and other styling motifs to appeal to mass markets. Detractors called their 'school of streamlining' vulgar and dishonest, but it appealed to consumers, especially women, and it undoubtedly played a major role in helping US industry to climb out of the Depression after the calamities brought on by the 1929 Wall Street Crash.

Raymond Loewy (1893-1986) epitomized this brave new world of commercial design with his war cry of 'la laideur se vend mal' (ugliness doesn't sell). Manufacturers were impressed by this flamboyant Franco-American showman who wanted to 'streamline the sales curve upwards'. Here was a designer speaking a language they understood. Loewy went on to redesign some of the most significant icons of American life: the Studdebaker car, Lucky Strike cigarette packaging, Coca Cola dispensers and the interior of President Kennedy's personal jet. By 1946 he had a roster of 75 international clients in a range of industries, and he was able to extend his influence in the post-war years with offices in London and Paris. Loewy's peers were no less influential. Henry Dreyfuss (1904-1972), for example, designed vacuum cleaners for Hoover, televisions for RCA, and the long-serving Bell telephone, as well as writing the classic text on ergonomics, *Designing for People.* Like Loewy, Dreyfuss built a consulting organization which is still in business and highly respected today.

The distinction of founding the world's first industrial design consulting office was claimed in 1926 by Walter Dorwin Teague (1883-1960). He established relationships with a number of large US corporations, starting with styling exercises for Kodak. But all claims and contracts of the period were fiercely contested by rival consultants, as a pattern was set which the modern design consultant will immediately recognize. The US pioneer designers were not shrinking artists but aggressive business people. They hustled for business, jealously guarded clients, and hired and fired design staff as contracts came and went. A community of draughts people and modelmakers developed around them to service the bigger jobs.

Visionary thinking coexisted with cut-throat business practice in these fledgling design firms. Nobody epitomized the bizarre juxtapositions more than Norman Bel Geddes (1893-1958) whose theatre background influenced his brand of futuristic design. Geddes brilliantly predicted an age of freeway systems and air-conditioning, and gave shape to streamlined buses and ocean liners, which were never built. But he also instructed his lawyer on one occasion, in dealing with a client, to 'get as much money out of them as you can'.

In New York in the 1930s a design industry with distinctive modern characteristics emerged. Highly motivated and competitive consultants charged companies large fees to devise new products by linking consumer psychology and advancing technology to the traditional design skills of thinking, drawing and visualizing. It was on precisely that basis that the design business was to boom in Britain 50 years later.

Britain can make it

The influence of Loewy and his contemporaries took a long time to make itself felt in Britain. In the 1930s, sleepy British industry with its large captive Empire markets was hostile to new styling trends from across the Atlantic, while graphic design was largely in the hands of printers, and retail design was the domain of shopfitters. Yet two far-sighted British designers, Milner Gray (born 1899) and Misha Black (1910-1977), had visited New York and seen what was happening.

After working on propaganda exhibitions for the Ministry of Information in the early 1940s, Gray and Black founded Britain's first modern design consultancy Design Research Unit (DRU) in 1943. They based their new firm on the American model, announcing that the aim of DRU was to 'present a service so complete that it could undertake any design case that might confront the State, Municipal Authorities, Industry or Commerce'.

Fitted kitchen designs by Milner Gray were shown at the 1946 *Britain Can Make It* exhibition, a flagwaving exercise which neatly modified the wartime 'Britain Can Take It' slogan to set the agenda for post-war British design. DRU subsequently went on to design a wide range of projects from Pyrex glass ovenware to British Rail's corporate identity, reflecting its multidisciplinary status as a consultancy. The business survives to this day, although it has been partially eclipsed by a great many other British design firms which owe much to DRU's pioneering work.

The struggle for recognition

After the 1951 *Festival of Britain*, a showcase which celebrated the centenary of the 1851 *Great Exhibition* and gave a boost to the avant garde ideas of young UK architects and designers, attention switched away from Britain to developments in Italy, Scandinavia and the US. In Italy, the period known as the *ricostruzione* placed industrial design in the vanguard of post-war social and industrial regeneration. Lighting, furniture, fashion and cars all embraced a national design ethic. Young architect-trained Italian design consultants such as Achille Castiglioni and Ettore Sottsass began producing work for enlightened family-run manufacturers. Their products were destined to make Italian style one of the most potent symbols in international markets.

In Scandinavia, a series of exhibitions in the 1950s promoted the concept of Scandinavian Modern, a more human and naturalistic expression of modern design than the brutal exercises in metal and concrete elsewhere in Europe. There was a recognizable link with the region's craft traditions, and an emphasis on natural materials and organic forms. Architects such as Alvar Aalto and Arne Jacobsen produced a range of industrial design products which greatly influenced British designers. In America, the post-war consumer boom was underway with chrome-laden and tail-finned automobiles designed by Harley Earl for General Motors, a symbol of a society in feverish pursuit of the American Dream.

Designers in Britain, meanwhile, were still struggling for recognition and unconnected to the mainstream of business life. They were known as commercial artists and were looked down upon by advertising agencies, which controlled most packaging and graphics work at the time, and by architectural firms, which designed interiors, exhibitions and furniture. British manufacturing was dominated by engineers and suspicious of what 'bohemian' industrial designers could offer. It wasn't until the 1960s that British designers could shake free from the austere pattern of the immediate post-war years, liberated by the explosion of new creative ideas which accompanied the emergence of the pop aesthetic. Suddenly 'Swinging London', not New York or Milan or Copenhagen, was at the centre of new visual trends.

The primacy of pop

In a freewheeling atmosphere of artistic ferment, British designers grew in confidence. Pop art, music and fashion propelled British design to centre stage and there was a feeling that anything was possible. It was in the 1960s that Wolff met Olins, Minale met Tattersfield, Conran worked with Fitch, and Fletcher, Forbes and Gill created the forerunner to Pentagram. Design firms established in this period became the bedrock of the modern British design business. Today's elder statesmen of design were ambitious and energized young entrepreneurs in the 1960s. They set their sights high but they had to endure the prolonged economic slump of the early 1970s before realizing their ambitions to the full in the boom years of the 1980s.

The global experiment of Unimark

By the1960s, companies were becoming multinational organizations with trading patterns which took them into new markets all over the world. Designers before Raymond Loewy had only looked to national, regional or local markets. Now a major opportunity arose to create a multinational design firm which could provide a truly international service for the biggest clients. Swiftest off the mark was Unimark International, a consultancy created in Chicago in 1965 by Ralph Eckerstrom. This firm quickly opened a string of offices in America, Europe and the Far East.

Eckerstrom, described by colleagues as a 'Kennedyesque figure', gathered designers of the calibre of Bob Noorda, Massimo Vigneli, Jay Doblin and Larry Klein around him. The Bauhaus luminary Herbert Bayer joined the Unimark board, providing an important link with the past. By 1969 Unimark appeared to be on course: the firm earned nearly a million dollars in fees in the first three months of that year. With 500 staff on the payroll, it did much pioneering work in the use of computers in design and in developing a more scientific and anthropological approach to design.

But in the early 1970s, Unimark's world fell apart: its biggest corporate identity client, Ford, went on strike, forcing the closure of its Detroit office. Another client, Memorex, froze all investment in Europe. Unimark's London office was forced to close even before Britain's three-day week. By 1973, Unimark's empire was in tatters, the victim of too rapid expansion and economic recession which saw design budgets slashed almost overnight. The lessons of Unimark were not learned by the young British designers who watched the saga unfold. A decade later, many of the same mistakes in multinational design were repeated by the British design industry.

Decoding the designer decade

The 1980s started slowly for British designers with the economic downturn of 1981-82 once more plunging fragile consultancies into gloom. But for UK design there was one important difference to the cyclical pattern of fortunes. Prime Minister Margaret Thatcher had decided to make design a central focus of her industrial and economic strategy and throughout the 1980s government money was available to 'prime the pump' for investment in design by private companies.

In retrospect two events provided turning points. First, in 1980, Allied International Designers, led by James Pilditch, became the first design consultancy to 'go public' on the Stock Exchange. The City was forced to take the design industry seriously: it could no longer be dismissed as a marginal cultural phenomenon and was fast becoming a wealth creator in its own right.

Then, in 1982, the Prime Minister held a Downing Street 'design summit' attended by 70 leading design figures. A plan of action was drawn up under the leadership of industry minister for design John Butcher, who regarded design as 'a competitive weapon in an industrial and economic war'. Government funding of design rose from £4 million in 1982 to nearly £14 million in 1988, at the peak of the boom. Exhibitions, seminars and publications channelled through the Design Council were aimed at encouraging British industrialists to employ designers. Part of this programme involved running a funded consultancy scheme which enabled British companies to claim a subsidy on the cost of hiring a design consultant.

Government support was important, but other factors also played their part in the creation of a design boom, both independently and cumulatively. Monetary policy relaxed credit controls, so leading to a rush of activity on the high street. Retail and packaging design commissions proliferated as companies rushed to grab a share of fast-growing consumer spending. Cheaper mortgages encouraged a house property boom, so stimulating the design-led sector of home furnishings. Office redevelopment and corporate communications similarly simmered to the boil as the construction and advertising industries blossomed.

An expanding client base

Wherever British graphic, product and environmental designers looked, the traditional client base was expanding. The Kodaks, Olivettis and IBMs, which were once the prime patrons of design consultancy, were suddenly joined in a service sector revolution by fast food chains, trade unions, public schools, theme park owners and all manner of new design users. International companies began to regard London as a design consultancy capital and placed commissions with British firms. From 1982 onwards, it is estimated that more than 70 per cent of all British consultancies did some work for overseas clients.

Sir Terence Conran carved out a large retail empire for his Storehouse combine, so integrating design with business on a scale hitherto unknown. And, in an era which promoted the virtues of self-help, enterprise and self-employment, many designers decided to run their own businesses. By the mid-1980s a large, sprawling and unwieldy industry was in existence, with as many as 2,000 different groups working in graphic, product, corporate and environmental design.

Some firms stayed small and craft-oriented. Others grew rapidly in size and sophistication: they expanded into lucrative non-design areas such as naming and branding, market research, management consultancy and production engineering; and they introduced computers into the design process to improve productivity.

Skilled staff from advertising, marketing and manufacturing were attracted by the entrepreneurial spirit and attractive pay and conditions of design consultancy. The entire industry snowballed rapidly as more and more design firms sought a Stock Market listing. With public money these larger groups then embarked on a Unimark-style expansion programme, with acquisitions in the US as a prime target in the bid to service the multinational client.

By 1988 the British design industry was employing more than 30,000 people and was worth nearly £2 billion. According to *Design Week*, the leading players had invested more than £70 million in the US design market through a series of takeovers. That year Britain's 100 largest design firms earned £450 million in fee income, a 50 per cent increase on the previous 12 months for the second year running.

Design was regarded as in tune with the cultural tenor of the time. This was the designer decade, an age of appearance, of conspicuous consumption. It didn't appear to matter that government money earmarked for manufacturing had ironically fuelled a boom in packaging and identity: a triumph of image over content.

Boom turns to bust

Even at its peak, design was not infallible. City analysts complained of poor management and financial controls within design consultancies. Clients complained of excessive fees and superficial designs which followed fashion blindly rather than solved real business problems. For all their popularity, designers had never emulated advertising agencies and succeeded in negotiating long-term retainers with clients. Nearly all work was on a project-by-project basis, even if some of the relationships with clients were well established. In 1987 Wally Olins reminded his colleagues that 'design is still largely a jobbing trade. Major companies still pick up designers and drop them in a way that they wouldn't dare with other professional advisers'.

Economic downturn at the end of the 1980s exposed the cracks in the facade of the British design business. The government's need to control inflation through the mechanism of high interest rates made borrowing more expensive. Consumers found they had less money to spend; retail and housing markets started to suffer. Designers, who had borrowed more than most to fund expansion of their businesses via international acquisitions, became more strapped for cash than most.

British design suffered traumatically in the recession at the start of the 1990s. There were spectacular casualties. Among the best known was the Michael Peters Group, a creative thoroughbred from the 1970s, which collapsed as a result of mounting debts following rapid international growth. At its peak, the Peters design and communications empire turned over £49 million and employed more than 700 staff in a network of offices from Madrid to Minneapolis. The job losses throughout UK design proved how closely the industry is linked to the advertising and construction industries. When marketing and building budgets begin to falter, designers are among the first to suffer the consequences.

Recession also dampened other encouraging trends such as the emergence of the professional design manager. Just as the rise of the marketing manager in the 1970s enabled advertising agencies to sell their wares to companies more effectively, so the rise of the design manager in the 1980s, most notably in retail and transport organizations, facilitated bigger and better commissions for design consultants.

Consolidating in the 1990s

During the mid-1980s the British design industry grew at a rate estimated at 30 per cent a year. One effect of recession was to slow that rate of growth practically to a standstill. The opportunity for the design industry to pause and catch its breath enabled professional consultants to reassess the way they conduct their business. There has been much soul-searching as a new agenda has gradually emerged for British design in the 1990s. That agenda involves a revival of social idealism, manifested in the interest shown by designers in such issues as public transport systems and environmental recycling. It also involves a renewed emphasis on the craft of designing, with its distinct cultural heritage.

There is recognition that British design became so much a part of the marketing mix in the 1980s that it lost its core identity amid the share flotations and global acquisitions, market surveys and communications audits. A common refrain was that 'it has lost its soul'. Neville Brody, one of the best-known graphic design talents of the 1980s, complained that 'design in England has become a commodity'.

Now there is interest in the business of design allied to the history of design within a cultural framework, a recognition that design is about art and also about the social mix in the 1990s. Public sector design, particularly in relation to the fabric and environment of cities, has become a major issue. Government departments, education institutions and local authorities are emerging as key users of design services.

So British design, tied to the notoriously unstable post-war UK economy, is locked into a cycle of change. Once again it is undergoing a profound and painful transition. However, most of the gains of the past decade have not been wiped out by economic downturn, and indeed they are considerable.

Forged from disparate roots

The British design business has the largest institutional and educational framework in the world, and a historical perspective which suggests that it has distilled the best qualities of the professions from which its disciplines emerged. Design in Britain is forged from disparate roots in art, craft, construction and marketing. It draws its:

- skill, vision and lateral thinking from art
- project management and social responsibility from architecture
- boldness of communication and spirit of entrepreneurship from advertising
- versatility in achieving business objectives from marketing.

The design industry is a hybrid but none the worse for that. Not surprisingly it earns its money is a variety of different ways by:

- charging a set fee for the project
- hiring out designers at a fixed hourly rate (most common among graphic designers)
- negotiating a royalty on sales (most common among product designers)

- taking a percentage of the total spend on furniture and materials for environments (most common among interior designers).

This last method of remuneration reflects the ad agency cut of total media spend.

The institutions which represent the design industry's interests are equally many and varied, often overlapping in functions and competing for the same pots of money from government, designers or sponsors.

The institutional framework

The Design Council

Founded in 1944 as the Council for Industrial Design, this government quango has always had the remit to 'promote by all practicable means the improvement of design in the products of British industry'. That does not necesarily mean promoting the British design industry, a major bone of contention for many designers who have felt the Design Council should do more to represent their specific interests.

The Council played a leading role in the 1946 *Britain Can Make It* exhibition and the 1951 *Festival of Britain.* Under the directorship of Gordon Russell in the 1950s it opened a London Design Centre in the West End and built a strong international reputation as many countries copied its model of operation. In 1972 it was renamed the Design Council and, under director Paul Reilly, continued to campaign for the social and economic benefits of good design. In the 1980s, under director Keith Grant, it enjoyed a central role in the government's efforts to give design a higher profile in industry. Recently the Council has modified its strategy in the face of a worsening trade deficit in manufactured goods. Under current director general Ivor Owen, it is concentrating resources on industrial and educational audiences. The Council has also targeted selected key industrial sectors such as furniture and textiles, rather than the broad spectrum of industry, in order to increase impact.

The Design Council has 250 staff and a turnover of £12 million, roughly half of which is provided in grant aid from the Department of Trade and Industry. Its award schemes, publications and committees continue the work begun in the 1940s. It is increasingly an uphill battle.

Chartered Society of Designers

Founded in 1930 as the Society of Industrial Artists (SIA) and later known as the Society of Industrial Artists and Designers (SIAD), the Chartered Society of Designers (CSD) is the largest professional organization of its kind in the world. It has around 9,000 members, working in a wide range of design disciplines, and is dedicated to speaking out on issues which affect them and upholding the professional standards of the individual designer. Its work on professional codes of practice and in design education is intended to protect the credibility of a profession which cannot operate a bar on entry. Anybody, trained or not, can call themselves a designer and set themselves up as one. Entry to the CSD is by inspection of a portfolio, and entitles members to use the professional affix MCSD (for Members) or FCSD (for Fellows). The Society has adopted plans to widen membership in certain categories to design managers and non-designers.

Design Business Association

Founded in 1986, this small and energetic body has emerged from within the ranks of the Chartered Society of Designers to address a specific purpose: to represent the commercial interests of design consultancies rather than reflect the professional values of the individual designer. It is a trade association with more than 200 design firms as members. It offers a range of business services and has launched a number of schemes to promote design including the DBA Design Effectiveness Awards, which judges work on the basis of commercial effectiveness rather than aesthetic merit. The DBA plays a major role in promoting the importance of design to business, not only in Britain but in the growing European market for UK design expertise.

RSA

Founded in 1754, the RSA (the Royal Society for the Encouragement of Arts, Manufactures and Commerce) is the oldest and most venerable of all design institutions. The RSA commands respect throughout the world. Its work also focuses on the environment, encouraging collaboration between architects, developers, artists and craftspeople. Its design section runs a programme dedicated to bringing young designers and industry closer together, most notably through a bursary scheme.

In 1936, the RSA founded the Faculty of the Royal Designers for Industry (RDIs), a prestigious association limited in numbers which confers the title of RDI on leading international designers.

Design Museum

Probably Britain's youngest national institution, the Design Museum opened its doors in London Docklands in 1989. Established with a £7-million gift to the nation from the Conran Foundation, the museum now looks to commercial income, sponsorship and government support to develop its international programme of lectures, displays and exhibitions. Its aim is to 'enable everybody to understand and appreciate the effect of design on the products, communications and environments we use', and the Museum does much pioneering work in education. Its forerunner, the Boilerhouse Project in the Victoria and Albert Museum, links the Design Museum directly back to the work of the great mid-nineteenth-century design reformer Henry Cole.

D&AD

Founded in London in 1962 by a group of advertising creative directors, art directors and designers as the Design & Art Directors' Club, this body exists to 'stimulate, not congratulate' high standards in design and advertising. Its publications and awards have been criticized as elitist, although their prestige is not in doubt. As D&AD is shared by ad people and designers, it has also highlighted the rifts between the two groups as the advertising and design industries have gone their separate ways.

There are many other institutions which impinge on the working life of the commercial designer, such as the Ergonomics Society, Design Research Society, Royal Institute of British Architects, Crafts Council and Marketing Society.

The educational scene

There is a large and complex framework of design teaching at further and higher education level. There are now an estimated 50,000 students enrolled on 1,000 different design or design-related courses in more than 300 colleges and polytechnics in the UK. Every year between 6,000 and 7,000 design graduates seek employment in the design profession. There was a dramatic expansion of courses and student numbers in the late 1980s in direct response to the demands of a design industry with full order books and bright prospects.

Many design consultants now consider that these figures are too high, and that the design industry cannot acommodate so many entrants. They want to see the design profession exercising control over student numbers in the way that the medical profession controls the study of medicine. However, educationists respond that design courses have broader educational and intellectual responsibilities and must not be regarded solely as vocational training.

Art versus business

The art-versus-business argument goes back to the art-and-industry debates of the mid-nineteenth century. As former Design Museum director Helen Rees recalls: 'A conflict broke out around the question which still plagues design education today: should the schools simply provide a training in wealth creation, or should they teach design as an adjunct of fine art, which was deemed to be both morally and culturally uplifting? In other words, was design simply at the service of commerce or should it have a broader social purpose?' The answer, says Rees, was a typical British fudge: an attempt to accomodate both views. Analysis of the modern development of both design education and practice reveals the same attempt to reconcile art and business, the commercial and the cultural. It is a delicate balance: when British designers went wholeheartedly commercial in the 1980s, perhaps losing sight of social values, they suffered when the economy deteriorated. In earlier decades when they clung to cultural and craft values, they found it difficult to exert a full influence on business life.

The challenge of the 1990s must be to make the business of design an effective commercial tool and a force for social and cultural improvement, so binding together the disparate strands of the design industry's development. Emerging trends such as the 'green' consumer and the 'ethical' company will help such a mission. So will the new design freedoms provided by quantum leaps in technology.

Utopianism and entrepreneurship are not mutually exclusive. If they had been, then Robert Adam, Christopher Dresser, Raymond Loewy, Gordon Russell and Terence Conran would never have made the mark they did on the heritage of British design.

Further reading

Aldersey-Williams, Hugh (1988) *New American Design: Products and Graphics for a Post-industrial Age.* New York: Rizzoli.

Bayley, Stephen (ed.) (1985) *Conran Directory of Design.* London: Conran Octopus.

Dormer, Peter (1990) *The Meanings of Modern Design.* London: Thames & Hudson.

Henrion, FHK (1983) *Top Graphic Design: Examples of Visual Communication by Leading Graphic Designers.* Zurich: ABC Editions.

Heskett, John (1980) *Industrial Design.* London: Thames & Hudson.

Jackson, Lesley (1991) *The New Look: Design in the Fifties.* London: Thames & Hudson.

Lucie-Smith, Edward (1983) *A History of Industrial Design.* London: Phaidon.

McDermott, Catherine (1987) *Street Style: British Design in the 1980s.* London: Design Council.

Myerson, Jeremy and Katz, Sylvia (1990) *Lamps and Lighting.* London: Conran Octopus.

Myerson, Jeremy and Katz, Sylvia (1990) *Home Office.* London: Conran Octopus.

Myerson, Jeremy and Katz, Sylvia (1990) *Kitchenware.* London: Conran Octopus .

Myerson, Jeremy and Katz, Sylvia (1990) *Tableware.* London: Conran Octopus.

Sparke, Penny (1987) *Design In Context: History, Application and Development of Design.* London: Bloomsbury.

Sparke, Penny et al (1986) *The Design Source Book.* London: MacDonald.

Keren Ludlow/Sharp Practice

'The impetus towards design in industrial life today must be considered from three viewpoints: the consumer's, the manufacturers's and the artist's. In his appreciation of the importance of design the artist is somewhat ahead of the consumer, while the average manufacturer is further behind the consumer than the artist. The viewpoint of each is rapidly changing, developing, fusing. More than that, the economic situation is stimulating an unanimity of emphasis, a merger of viewpoints.'

Norman Bel Geddes
Horizons

2 PERCEPTIONS OF DESIGN
Eric Schneider

It was to be some 50 years after the decade in which Norman Bel Geddes and his contemporaries streamlined America, that the UK design sector finally came of age. By the end of the 1980s high streets had been transformed, company identities metamorphosed, brands revamped and business perception of design shifted from the domain of the effete into a valid concern for even the most hard-headed manager. Within this context this chapter presents a set of management-based perceptions on design and its worth, with a view to readers understanding and questioning the role of design and the links between design and management. The chapter also considers why, despite the boom, some argue that design remains underutilized as a corporate resource, its influence no more than skin deep, and what key challenges face the sector as it matures. The chapter is organized around ten assertions.

1 Design must be seen in context

The US and UK experience of design poses two key questions: why did it take so long for design to catch on in the UK, and, why did design fall in prominence in the US? Design exists in a social, historical, economic and political context which establishes constraints, perceptions and conditions for growth. To survive, design must meet the needs and challenges of that context. The danger is to assume that an interest in and commitment to design is self-evident and automatic.

As highlighted in chapter 1, there was nothing new in the UK recognition of the need to improve design. Such awareness, and the battle to integrate art with science, and design with business, dates back to the founding of the South Kensington museums and the establishment of a Royal Commission to investigate the relative design failings of UK companies at the Paris *World Exhibition* of 1867. Until the 1980s the UK appeared to lack the right combination of economic urgency, sympathetic government, cultural and educational factors to commit itself to a design policy. As in the US, where design grew out of depression as advertising agencies and their clients resorted to artists and stylists to reduce 'consumer drag' and boost the flow of sales, adverse economic and competitive factors provided the breeding ground for design in the UK. Government and private sector support for design grew out of a context of relative economic decline, a slowdown in global economic growth, high international competition, and increasingly fragmented and maturing markets where design was seen as a key factor for improving non-price competitiveness.

Although economic factors may provide the urgency for design, social and cultural factors can act as a major constraint. In the UK such factors have long worked against the effective application of design for business purposes. Our education system has traditionally separated the making of things (art and crafts) from the academic study of the arts, as well as from science. Similarly, business and trade have not been the culturally approved routes for the best graduates, nor for those seeking social recognition and status. Visual literacy, product orientation and problem-solving skills have tended to be undervalued in UK companies, particularly in comparison to the skills provided by legal and accounting professionals. Few designers and engineers reach board level, and fewer still are in positions of political influence and power.

2 Design must add value to be of worth

There is no reason why anyone should invest in design unless the perceived returns of that investment are believed to exceed the cost. Design is not art. This raises two issues. How does design add or create value, and how is that value to be measured? Both issues are often less than fully understood by design practitioners and users alike.

Broadly speaking, the areas of design commissioned by organizations consist of a mix of product, communication, environmental and corporate or brand identity design. These four areas may then in principle affect most aspects of a company's resources and activities, from staff working conditions to a large part of its production, promotion and marketing activities. The design of products, for example, may lead to competitive advantage by permitting a price premium through improved product attributes, reducing costs via production efficiencies, or improving responsiveness through shorter development and lead times. The design of corporate communications may result in cost savings through improved efficiencies (eg better form design for an insurance company) or greater marketing effectiveness. The (re)design of an office may enhance capital values, as well as physically improve working conditions and staff morale. And corporate and brand identity design may improve perceptions and loyalty as well as enhancing awareness and profile.

The difficulty is measuring and isolating the effectiveness of design performance. Two obstacles hamper progress. The first is that the effects of design may be any combination of: direct and indirect (or knock on); short and long term; tangible (eg measurable or physical) and intangible (eg non-measurable, perceptual, or emotive). The short-term, direct and tangible effects of design may be relatively straightforward to monitor, and can be measured in terms of the balance sheet (eg in terms of sales, costs, profits, stock values selling and administrative expenses, fixed asset values), share prices, and the accounting ratios such as return on investment which typically provide the main short-term criteria for performance for companies. However, the effect of design on brand values and goodwill is harder to quantify. And to measure the longer-term, indirect, effects of design is more difficult still.

The second obstacle is that when used effectively, design is likely to be highly integrated with other activities, thereby making it difficult, and maybe irrelevant, to isolate its effects from those of other functions such as marketing or production engineering. Moreover, although the visual aspects of design are of primary importance, design may also make significant non-visual or 'secondary' contributions. These may include the particular role, influence and/or skills of designers as individuals, as well as the culture associated with design (eg an overriding concern for artefacts and a problem-solving approach). Design may also have a key role as an integrating mechanism, providing a common sense of purpose and direction across the various business functions as well as assisting greater consistency in the actions, behaviour and outputs of companies. Finally, design may play an important role as a control tool and/or a barometer of organizational performance, reflecting the degree of control a company has over those outputs. Such non-visual areas of design contribution are likely to be increasingly significant as the ability to perform imaginatively and entrepreneurially becomes an ever more critical factor for organizational success.

For these reasons, a strategic approach to design should be based on an understanding of how design works within the 'value chain' (Schneider 1989) of a business: how it may impact in terms of the various interrelated activities of designing, producing, marketing and distributing products or services as the company adds value to its various inputs in order to make a profit and survive. Given the inevitable constraints on resources, such analysis should help identify where design may best contribute to achieving a competitive advantage, and where the thrust of design energies and investment can most effectively be applied.

3 Design is a word with different meanings

Most designers take it for granted that the term 'design' is understood. In practice it is used loosely to cover a breadth of meanings and activities, thereby providing scope for confusion. One example is in identifying the type of designer required, since many areas of design (eg aviation engineering, textile design, architecture, furniture design) are covered by the one term, although the practice involved may vary considerably in terms of complexity of application and specialization. Even within the area of graphic design, narrowly differentiated functional subdivisions (such as distinguishing packaging from print, point-of-sale from company literature, and corporate identity from corporate reports) mean that managers can be uncertain about the nature of their needs and the type of designer to use.

There is also scope for misunderstanding the meaning of design. The *Concise Oxford Dictionary* lists over 20 definitions. In his book, *Design for Corporate Culture*, Fairhead (1988) uses the sentence 'the success of Japanese design firms is largely due to good design' to explain how the word 'design' may refer to the end product of design (eg hifi goods), one particular functional activity (eg industrial design) and/or to designing (ie the design process). Peter Gorb's definition (Gorb 1983) is precise: 'design is the planning process for artefacts'. This clarifies design as being concerned with the whole process, from conception to final production. It also emphasizes the management role in relation to design activities, and provides a basis for evaluating design in relation to objectives, whilst presenting design as an interdisciplinary activity, not just exclusively the realm of professional designers.

4 Designers are not the only designers

This point is critical. Although there are some areas where a designer-craftsperson may be responsible for an idea and its implementation (say in developing a one-off chair), this is increasingly rare.

In most organizations, design is a team-based activity. For example, the successful design and development of a new product is likely to involve a whole range of cross-functional business activities and personnel, from marketing to finance, production to sales, and design to distribution. Angela Dumas (Gorb and Dumas 1986) has introduced the term 'silent designers' for non-design professionals who play key roles in the design process, and in determining the success or failure of a design project.

5 Successful design requires much non-design work

The successful application of design is likely to require an appropriate mix of supporting factors. The best results may be expected when there is congruence, or fit, between the needs of design and the supporting factors which may include management style and leadership (eg top-level support, sponsors and protectors); staff (eg with appropriate allocated responsibilities, authority, motivation); skills (achieved through appropriate training and experience); structure (eg in terms of the location of design function, team composition etc); shared values/culture (eg in terms of attitudes to design and to products); systems and procedures (eg for communications and control) and strategies.

But there is no simple all-embracing solution. The conditions for achieving the successful design of a safe power station are likely to differ from those required to create a radically new electronics product, a landscape garden or a piece of graphics. Furthermore, those needs and conditions may vary again depending on whether a project is at the problem analysis stage, the creative ideas development stage, or the stage of implementation and/or construction.

In short, for success the overall process requires skilful management and integration of activities.

6 Design management matters

It is now easy to understand how the discipline of design management developed in the 1980s and to appreciate the interest and faith placed in it, at least by academics, designers and some companies.

Not only in its very title did design management imply that it could resolve the difficult issue of how best to manage creativity; in one phrase it also bridged, integrated and seamlessly merged two cultures previously divided by an enormous gap. In addition it epitomized the new commercialism and business orientation which characterized the UK design sector of the 1980s.

In practice, however, the term 'design management' may be as confusing as 'design', being commonly used to refer to at least five main areas of activity, including educating managers about design; educating designers about management; managing design practices; managing projects; managing design as a corporate resource or function, at both strategical and tactical levels.

Moreover, as a young discipline, with its boundaries yet to be established, and its foundations yet to be formalized, design management has been an attractive bandwagon for freeloaders. Many claims have been made for and on behalf of the subject which cannot be justified. Although much can be hypothesized on the basis of adjacent areas of study (eg from research into innovative cultures and organizations, new product development), considerable additional work is required to establish comprehensive management guidelines for design.

None of this is to decry the underlying aims and ambitions of the study of design management, which remain honourable. Sustainable design success requires the effective integration and management of design within an organization, as with the strategic application of any other business activity.

However, there remains a need for caution and restraint in making claims for a discipline which is still developing and has yet to gain the support of an appropriate body of academic research.

7 Professionalism pays

It is worth examining the rise of interest in management issues amongst designers, when the topic was until recently near anathema for many 'creatives'. One interpretation is that management can be expected to develop in direct response to a rise in the costs of failing to manage effectively. In the architectural profession the threat of litigation from clients and builders meant that long ago the development of project and contract-based management skills became integral to the architect's training. In other areas of design such pressures have been less and it has tended to be competitive forces which have been behind the trend for consultancies both to provide management skills as part of their service (especially in corporate identity design) and, as firms have grown and the economic climate toughened, to seek means to more effectively manage and control their own businesses.

Ironically, if designers, driven by commercial motives, have sought to increase their 'professionalism' in order to be taken seriously by clients, the trend in the architectural profession resulting from both legislative and market factors has been away from purely professional to more commercial values, reflected in a greater interest in marketing and business issues. Some would argue that the cost in architectural terms is already evident, pointing to how a concern for public urban space has diminished in the face of the commercial needs of the private sector. They may also suggest that increased commercial orientation can erode core professional values, as relationships governed primarily by fees and commercial gain replace those based on trust and professional standards, and as a shorter-term orientation leads to immediate client needs dominating wider concerns.

With broader societal and environmental issues impinging on all our lives, designers may find themselves facing the challenge of maintaining and improving professional standards, whilst meeting the requirements of commercial success.

8 Marketing myopia is dangerous

The growth of design industry expertise in addressing the marketing needs of clients has been another key trend. In principle, the desire to satisfy consumer (and societal) needs profitably is admirable, and the selection of target markets, the identification of key points of influence, and the development of products and services to satisfy market needs entirely logical.

But is a marketing orientation always wise in practice? In 1980 Wally Olins (see Olins 1980) linked the emergence of marketing in British business with the decline of UK industry as a major force in world markets. He was making the point that marketing which fails to understand the core identity of a business often leads to short sighted 'me-too-ism' based on the latest perceived fashions and trends. Other writers have also made the case that sometimes the dominance of marketing may actually have a negative effect on organizational performance.

Two arguments prevail. The first is that one of the key contributions made by designers is their imaginative approach. One danger is that despite the quest for imaginative marketing, often in practice such imagination falls victim to rational analysis and linear thinking. Second, over-efficient responses to marketing needs may cause designers to shoot themselves in the foot, especially those seeking to differentiate their consultancies on the basis of strong ideas and/or visual design. Many highly professional service-oriented architectural practices, who won repeat commissions by designing buildings which satisfied their unimaginative clients, now find themselves identified with work of this type rather than quality architectural design. Similarly, some elements of the design profession in the US may have become too geared to corporate needs, too predictable in their output.

Used properly marketing can assist the designer both to improve his/her own commercial success by better understanding design consultancy market needs and, by understanding those of the client, to provide a more effective service. An understanding of market 'pull' forces must be balanced by an assessment of the resources and needs of the consultancy. And in a highly competitive situation marketing must be imaginative. It is for such creativity, particularly its visual manifestations, that clients come to designers rather than accountants or marketing consultants. And it is such creativity that differentiates designers from one another.

The challenge for designers is to use marketing to support creativity and not to let their core business be destroyed through poorly applied marketing techniques.

9 There are no free lunches

It is little surprise then that when a reverse in economic fortunes hit the design sector after a decade of spectacular performance and growth it was accompanied by cries of disillusionment with marketing and strategy from some corners. 'Back to the roots', was the call. But, arguably, the writing was on the wall. The design sector was highly competitive, offered prospects of excellent growth and margins, and was characterized by few barriers to entry. Even without the effects of economic recession, after a decade of growth such factors could have been expected to lead to a continuing stream of new entrants and to the eventual 'maturing' of the sector.

The typical characteristics of maturation include over-capacity; declining growth; intensified competition; increasing marketing activities; deals between players; a scramble by competing firms to find niches; high competition on fees; an erosion of profits leading eventually to a shake-out period. Combine this with a decline in clients' purchasing power, the availability of new DTP technologies and increasing competition both from adjacent sectors and internationally, and many of the factors characterizing the situation of the early 1990s fall into place.

Some would argue that with success designers became complacent. They oversold themselves, began to believe their own myths, and ignored the fact that in highly competitive markets gains from superficial design improvements are not sufficient to provide clients with long-term sustainable competitive advantage. Few asked whether design was delivering its promise, or sought evidence of genuine long-term success through design. Too few noticed that if the design boom was affecting the retail sector and fast-moving consumer goods (fmcg), it was also primarily about styling and communications and was leaving the manufacturing and heavy engineering sectors untouched and unmoved. Few clients seemed to observe that the real successes appeared to be in organizations (all too often Japanese) where design was integrated, and where managing design was taken as a normal and entirely necessary activity.

There are no free lunches, least of all in highly competitive markets.

10 It is hard to be green about design

It is now hard to be unaware of design, especially in consumer markets. Consumers, clients, competitors, staff and would-be entrants, have all become increasingly sophisticated. To survive in this climate designers must keep on their toes, remain competitive, identify niche market segments in which to differentiate themselves, and provide services and products in a manner that fulfils their promises and the needs of both clients and society.

Perhaps the major challenge facing the sector, however, will be the green issue. The economics of the sector mean that designers cannot survive by doing a job well just once. The very existence and growth of design businesses ultimately relies on continuing commissions and repeat business, on a pattern of continuing introduction of new and more designs. In the short term, of course, designers may assist their clients to supply ecologically and environmentally sound products. But the sector's lifeblood is the constant dissatisfaction with existing design solutions, and continuing consumerism. Whether design is sustainable in the long term in a situation of limited resources will most likely depend on how technology develops. Get ready for down-the-line mailers, re-programmable packaging, virtual reality interiors, and the other developments that designer technologists have in store.

Conclusion

This chapter has highlighted a series of management-based perspectives on design. It has endeavoured to highlight (1) that design must be understood in a context which provides both challenges and constraints; (2) that design has no relevance unless it adds value, and that value may not only be visual; and (3) that design should be considered as a process, rather than just the end result. It has also sought to emphasize (4) that design is a team-based activity, and that critical design decisions are also made by 'non-designers'; (5) that for design to work effectively requires an appropriate set of conditions; and (6) that how design is managed is of vital importance. Professionalism is important (7) but such values may suffer in the face of commercialism. And (8) although a marketing orientation is relevant, a blinkered and short-term marketing approach may be dangerous. Finally the chapter emphasized (9) that design is a highly competitive, and now maturing, sector in which survival will only be for the fittest. In the long term (10), the very nature of the sector may have to change radically if it is to fit in a world constrained by limited resources.

Further reading

Fairhead, J (1988) *Design for Corporate Culture: How to Build a Design and Innovation Culture.* London: National Economic Development Office.

Gorb, P (1983) Design and organisational outcomes. In Langdon, Richard and Cross, Nigel (eds) (1984) *Design Policy.* London: Design Council.

Gorb, P and Dumas, A (1986) Silent Design: An Interim Report of Research into the Organisational Place of Design in Industry. *London Business School Discussion Paper.* General Series GS-20-86. December.

Olins, W (1980) Where marketing failed. *Management Today.*

Schneider, E (1989) Unchaining the value of design. *European Management Journal* 7 (3) September: 320-31.

Paul Leith/Sharp Practice

'But to live outside the law, you must be honest.'

Bob Dylan
Absolutely* Sweet *Marie

3 THE LEGAL FRAMEWORK

Henry Lydiate

The function of this chapter is to help designers understand when, how and why to avoid legal pitfalls and seek professional legal advice. It explores the law and its relevance to the life of a well-run design consultancy, operating competently and successfully within the law, particularly as it applies to practising in the UK.

Legal terms are used only where necessary and always with a non-legal gloss. Whilst legal accuracy is guaranteed (at the time of writing), the material is no substitute for individual professional legal advice on particular problems arising in practice.

Designers and design practice

Everybody knows and is presumed by the law to know at least one legal maxim: ignorance of the law is no defence. Few bodies realize its permanent impact on their personal and professional lives.

Designers and design practices are bodies (organic and inorganic respectively) and both are therefore presumed to know the law as it affects their work.

Whether they practise in England, the UK, the EC, Australasia, or from a space station in the cosmos, designers and design consultancies need to know how to operate in a legal manner. Ignorance is no defence.

UK versus international laws

Design functions chiefly through the medium of the eye, which enables its products and services to be used and judged across language barriers with more ease than any other medium. It therefore has the ability to transcend international boundaries without translation, interpretation or other interference.

Modern technology, intensively and extensively used by all businesses for instant global audio and visual communication, puts design in the vanguard for attacking international markets.

Design deals containing a foreign (beyond the UK) element require that the practitioner knows enough about the laws of that foreign country to guarantee that such deals are legal at home and abroad. UK laws are very different to those abroad, and all foreign laws are not the same.

In the UK, knowledge and understanding of the legal framework within which design operates is the best starting-point for contemplating practising at home and abroad. Competence on the home front makes it possible to look outside at the legal requirements in countries where clients and deals may arise; the laws will be different in detail, but the broad principles will probably be similar.

The legal system: civil law

Most developed countries split their legal systems into two: civil and criminal. Both are relevant to design and design practice, in the UK and elsewhere. Courts and Judges in England and Wales for both systems are illustrated opposite. Scotland and Northern Ireland have their own system, personnel and proceedings, but the law is generally the same.

Proceedings in the civil system means that some body with a grievance against some other body has asked a civil court for a legal remedy to be given. Civil proceedings started in the UK cause the court to send a written notice to the alleged wrong-doer stating who is complaining, about what, the law allegedly breached, and an invitation to register a response by a stated date. From that stage to a final court hearing, many interim steps are taken which give each party opportunity to settle their differences without a court's order.

This process can and does take months, often years in a complex dispute, and is expensive. The court's costs are nominal, but legal fees often exceed the amount of money in dispute. In England and Wales, the plaintiff is the aggrieved body, the defendant the alleged wrong-doer, with the judge acting as referee and awarder of penalties in the form of judgement and orders.

Disobedience of a court's order can end in the defaulter's committal to prison and bankruptcy. Civil proceedings should be avoided at all costs, literally.

Nearly all civil proceedings are disputes over money: non-payment of invoices; failure to deliver goods and services; defects in designs or products of design; wrong and careless advice given and acted upon. The design business has all of these, but also further special legal areas: copyright; design right; moral rights; breach of confidence; and passing off (see chapter 15).

Whatever is in dispute, if the amount exceeds £25,000 the proceedings will be taken in the High Court in London and if below that sum in a local County Court. Appeal to the next higher court is possible, ultimately arriving at the House of Lords (the Law Lords sitting as a Judicial Committee or Court).

Evidence in civil cases is presented to the judge by the parties and is usually written, so as to clarify areas of dispute and agreements and to save everyone time and money. Disputes are decided by the judge weighing the evidence and legal arguments to find whether the plaintiff has proved the case.

The measure the judge uses can be put numerically and crudely at more than 50 per cent: is it more likely than unlikely that the plaintiff is right, in law and in fact? This balancing trick is described by the law as 'the balance of probabilities'.

The legal system: criminal law

Proceedings in the criminal system means that somebody is suspected of having done something that the law states is punishable with a fine or imprisonment, and requires the suspect to be brought before a criminal court to be tried, and if found guilty to be punished on behalf of the community.

This process can involve an arrest or in minor matters a summons to appear at a Magistrates' Court. Prosecutions may be conducted by private bodies including individuals, but are usually done by the Crown Prosecution Service at the instigation of the police. The court decides the guilt or non-guilt of the defendant (in the UK there is no verdict of 'innocent').

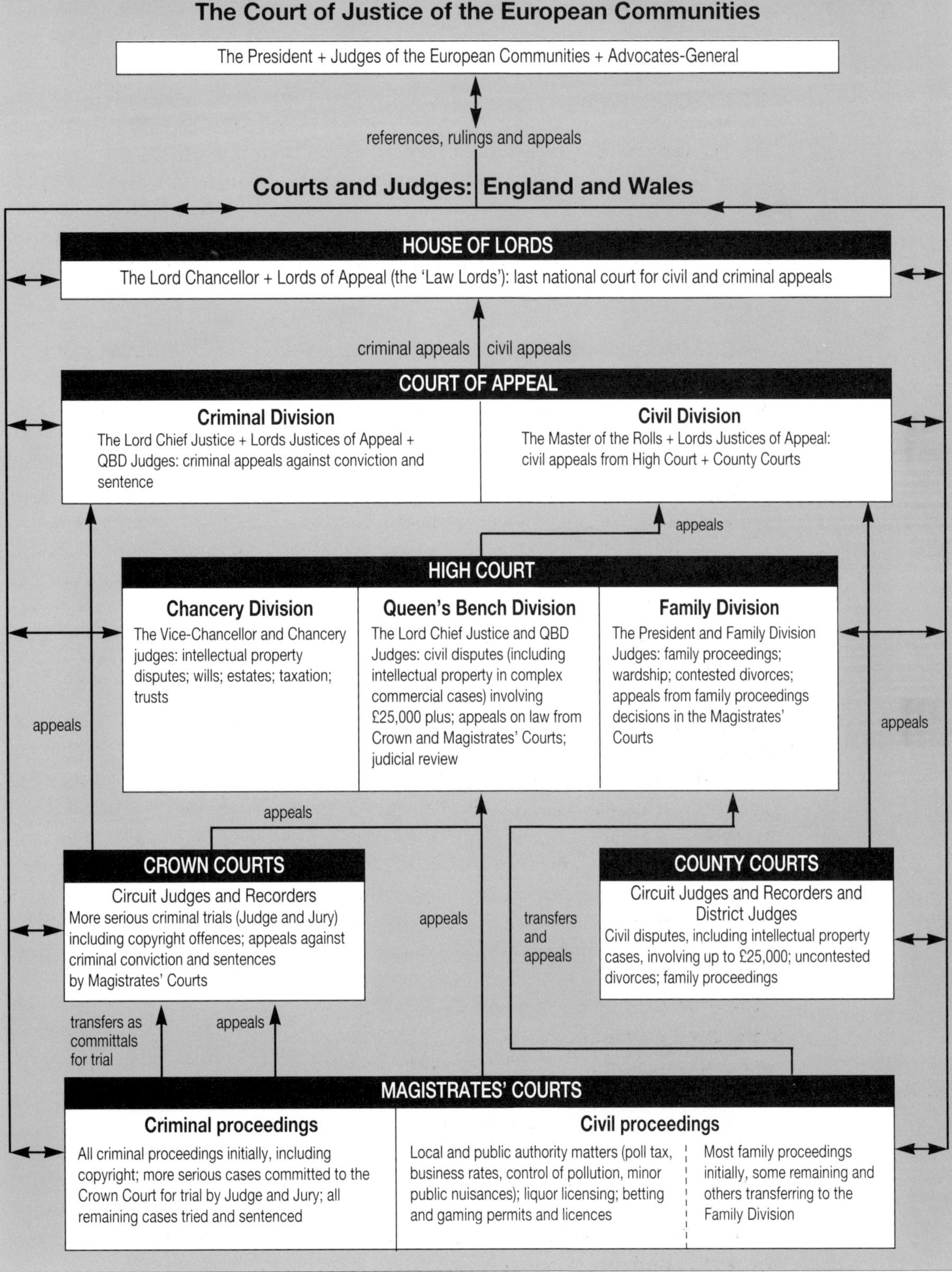
Courts and Judges: European Community and England & Wales
The Court of Justice of the European Communities
The President + Judges of the European Communities + Advocates-General
references, rulings and appeals
Courts and Judges: England and Wales
HOUSE OF LORDS
The Lord Chancellor + Lords of Appeal (the 'Law Lords'): last national court for civil and criminal appeals
criminal appeals
civil appeals
COURT OF APPEAL
Criminal Division
The Lord Chief Justice + Lords Justices of Appeal + QBD Judges: criminal appeals against conviction and sentence
Civil Division
The Master of the Rolls + Lords Justices of Appeal: civil appeals from High Court + County Courts
appeals
HIGH COURT
Chancery Division
The Vice-Chancellor and Chancery judges: intellectual property disputes; wills; estates; taxation; trusts
Queen's Bench Division
The Lord Chief Justice and QBD Judges: civil disputes (including intellectual property in complex commercial cases) involving £25,000 plus; appeals on law from Crown and Magistrates' Courts; judicial review
Family Division
The President and Family Division Judges: family proceedings; wardship; contested divorces; appeals from family proceedings decisions in the Magistrates' Courts
appeals
appeals
appeals
CROWN COURTS
Circuit Judges and Recorders
More serious criminal trials (Judge and Jury) including copyright offences; appeals against criminal conviction and sentences by Magistrates' Courts
appeals
transfers and appeals
COUNTY COURTS
Circuit Judges and Recorders and District Judges
Civil disputes, including intellectual property cases, involving up to £25,000; uncontested divorces; family proceedings
transfers as committals for trial
appeals
MAGISTRATES' COURTS
Criminal proceedings
All criminal proceedings initially, including copyright; more serious cases committed to the Crown Court for trial by Judge and Jury; all remaining cases tried and sentenced
Civil proceedings
Local and public authority matters (poll tax, business rates, control of pollution, minor public nuisances); liquor licensing; betting and gaming permits and licences
Most family proceedings initially, some remaining and others transferring to the Family Division

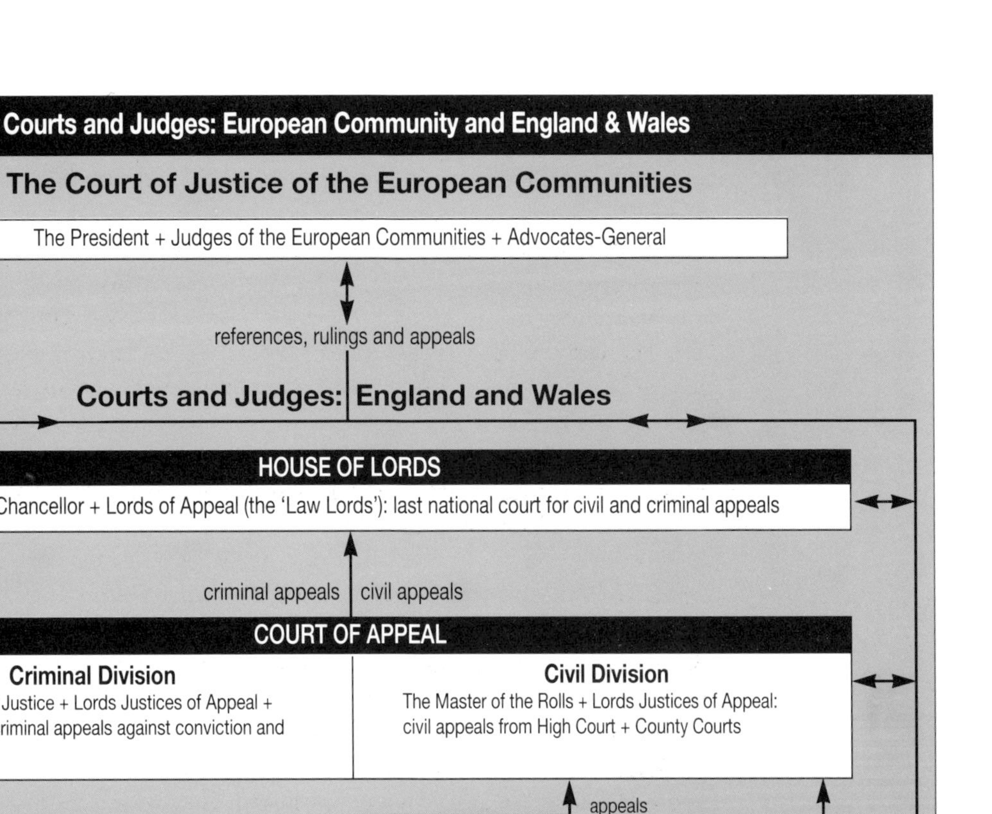

Ninety-five per cent of all criminal proceedings are completed in Magistrates' Courts locally, with the balance being sent to regional Crown Courts for trial by judge and jury. Design consultancy involvement with criminal courts could range from road traffic offences to copyright infringements. Appeal to the next higher court is possible, and ultimately to the House of Lords (as with civil cases).

Evidence in criminal cases is usually given orally, occasionally with documentation. The measure used by juries and magistrates to convict or acquit is whether the prosecution has proved its case through the evidence 'beyond reasonable doubt'. The 'burden of proof' is on the prosecutor to satisfy the jury or magistrates so that they are sure (not 100 per cent, and not beyond a shadow of doubt, but certain). This links to the message running throughout this book on the importance of making clear written agreements, and documenting project development carefully at all times.

Lawyers

Beyond the UK, most legal systems make no formal distinction between lawyers qualified to practise as advocates before the courts, and those qualified to give advice and assistance: attorneys, advocates, avocats, lawyers, notaires, are all names used abroad to describe a legally qualified professional. In the UK there is a formal distinction: barristers and solicitors, authorized to practise by separate legal bodies.

Barristers

Barristers are specialists in advocacy at court. They have rights of audience before every court in the UK, and exclusive rights in all but the County and Magistrates' Courts. Direct access by the public to barristers' services is forbidden (although certain professions, such as architects and accountants may approach them directly); only solicitors can commission and pay for their services as independent advocates.

Solicitors

Solicitors are specialists in advising and assisting the public. They have the exclusive right of direct access by the public, and limited rights of audience as advocates in the Magistrates' and County Courts. If a barrister is required for advocacy in the higher courts, the solicitor hires, briefs and pays for those services; barristers' fees are then put into solicitors' bills to their own clients.

Judiciary

The qualifications for judicial appointment are specified by an Act of Parliament, and barristers and solicitors are eligible for appointment. The higher the court, the higher the level of experience required; this is seven years in the lowest courts. Non-legal (lay) magistrates are by definition the only exception, but they receive compulsory training, before sitting and throughout their active service.

Costs: civil proceedings

Civil proceedings are often described as 'the ultimate game of chance'. The gamble is not so much in winning or losing at court, but in possessing the financial and emotional wherewithal to 'stay in the game' to the end. Civil litigation can be emotionally harrowing and financially bankrupting, win, lose, or draw. Costs of civil proceedings are the real problem. Plaintiffs must fund their own solicitor and barrister; if they win, the court can order the defendant to pay the plaintiff's costs, but might not always do so. Defendants fund their own solicitor and barrister; if they win, the court can, but need not, order the plaintiff to pay the defendant's costs.

There are many reasons why the 'winner' may fail to get an order for their costs to be paid by the 'loser'; but it is mainly because of money being paid into court at an early stage in the proceedings as an offer of settlement, and the offer being rejected by the other party.

For example:

- a designer sues a client for non-payment of a £30,000 fee for services rendered, which the client disputes and eventually offers to settle by paying £26,000 into court
- the designer rejects the offer; the money is in the court's bank account accruing interest from the date of deposit until the final trial date one year later, say
- the judge finds in favour of the designer, but only for payment of a £26,000 fee, just as the client had finally offered
- because the designer should, albeit with hindsight, have accepted the £26,000 and discontinued the proceedings, the judge probably would not order the defending client to pay the plaintiff designer's legal costs incurred over the twelve months dating from the payment of £26,000 into court.

The aim of this complex area of civil law is straightforward: to encourage both parties to settle their differences between themselves.

In practice, the vast majority of civil disputes are 'settled out of court', often 'at the courtroom door' on the first day of the trial, or even during the trial, as happened in the case brought by the Beatles' Apple Corps against the Apple computer company. Over 100 days were spent in court before the parties settled and asked the judge to make 'no order' as to costs (estimated at around £7 million).

'Men and women of straw', that is those without the means to pay what might be ordered, are a serious concern for plaintiffs; not only at the onset of legal proceedings, but also at the initial stages of contractual negotiations that might lead to a commercial deal.

Good initial legal advice will always involve an assessment of the means of the potential defendant, whether or not there is a good case in law.

Designers should take great care to assess the financial worth of prospective clients, both corporate and individual, before making a contract to provide services (see chapter 12).

Costs: criminal proceedings

Since criminal prosecutions are conducted on behalf of the public, all costs are largely borne by taxpayers from 'central funds' earmarked accordingly. Defendants acquitted receive their legal costs from central funds. Public prosecutors are paid by such funds, as are witnesses.

Convicted defendants might be ordered to pay a contribution to costs of the prosecution, but only if they have the means, and most defendants receive legal aid in serious cases (see below).

Free advice and assistance

The statutory Legal Aid Scheme, which funds individuals to defend or bring civil proceedings, is not available to businesses. Design consultancies may therefore find themselves suing or being sued by individuals who are eligible for legal aid, and having to fund their action or their defence themselves.

The Design Business Association publishes a series of leaflets, *Design & the Law*, which gives helpful pointers and consciousness-raisers; it also offers Professional Practice courses to which this publication is a complement.

The Chartered Society of Designers publishes a booklet, *Protecting Your Designs*, and offers guidance to its members on copyright and related matters. It also provides members with suggestions for standard forms of agreement, business terms and conditions, and conducts occasional seminars for members in which legal issues are sometimes covered.

Specialist bodies, such as the Association of Illustrators, and the Design and Artists Copyright Society are able to provide guidance on legal issues and trade practice in their own activity areas.

Forms of trading recognized by law

The freedom to trade given by the law requires that design practices establish themselves in a legal shape and form recognized and approved by law.

Consultancies reaching a stage in their growth which involves establishing a base, satellite or sister studio outside the UK, must take expert legal advice before doing so. This also applies to the EC, though to a lesser extent after 1992 when a good deal of harmonization of legal requirements will have been put in place throughout the member states of the Community.

Within the UK, there are several different forms of trading recognized by the law under which design consultancies may choose to operate.

Sole traders

Any adult can trade as a solo freelance designer using any name or logo they wish. No official permission is required.

Any profits made belong to the sole trader personally, subject to income tax and VAT. Any losses made are likewise the personal liability of the sole trader, including personal capital and assets owned outside the business activity. Therefore, the risks as well as the gains are entirely personal.

Partnerships

Two or more adults working together with the ultimate aim of making profits (even though they might fail) are recognized by the law as a partnership. Frequently, partnerships arise unintentionally when sole traders find themselves working together on one or more projects.

However, prospective partners often decide to work together and, before doing so, negotiate terms and conditions of their relationship including a business plan with outside financial backing.

It is advisable to have a partnership agreement drawn up by a solicitor and signed by the new partners. Whether or not such a legal document is written and signed, the law will recognize a working relationship aimed at profit-making and sharing to be a partnership, and will settle disputes between the partners, or the partnership and outside clients, according to statutory and common law provisions.

Essentially, these require that profits or losses are shared equally amongst the partners, unless there is clear evidence of a partnership agreement specifying different shares; but any debts owed by the partnership to outsiders are the liability of each partner (who then has to sue the other partners for repayment of their share of such losses). Each partner is therefore personally liable for all the partnership's trading losses, and this includes personal capital and assets owned beyond the firm's business activities. A partner's profits are subject to income tax and VAT.

Limited liability companies

Acts of Parliament provide for the creation of legally recognized and registered bodies, which have all the freedoms to trade and make profits or losses that are automatically available to individuals. These bodies are called companies, and are formed when two or more would-be company directors follow the registration procedures prescribed by company law legislation.The process is simple and cheap, a solicitor usually being used to guarantee accurate compliance, but charging modest fees, since company formation is legally straightforward.

The most important features of company law are that the directors can share the profits as they see fit, and register the limit of their personal legal liability to any sum they want to pay.

The Registrar of Companies makes available to the public all the company's and directors' financial and personal details.

Company accounts must be independently audited annually and filed with the Registrar for public inspection, and with the Inland Revenue for Corporation Tax purposes. Directors' meetings must be held and minuted, and corporate identity must include the company's registration number and the word 'Limited' or 'Ltd' after the company's name.

The most important feature of companies is that the directors are legally responsible for ensuring that trading does not at any stage cause the company to be unable to pay all its debts.

Any failure to prevent the company from 'trading unlawfully or insolvently', can make the directors personally liable to pay, despite their registered limited liability, and they could be disqualified from future directorships.

Creditors can sue for bad debts and may choose to bring 'winding-up' proceedings seeking a court order to close the company and sell its assets to clear its debts.

- ***Private limited companies*** are closed to the public who cannot buy shares and become directors.
- ***Public limited companies*** are open to the general public who may buy shares and become directors. Their corporate identity must include the words 'Public Limited Company' or 'PLC'/'plc'.

In practice, design consultancies tend to start as sole traders, develop into partnerships, incorporate as private companies, and perhaps transform into PLCs; but the reverse is also possible, sometimes out of choice, and frequently of necessity.

Other legal structures

Clubs, societies and associations can be formed by two or more individuals grouping together for mutual benefit, formally and informally. The law recognizes such free associations and provides that all the members are personally responsible for any group losses or debts.

Trade Unions have been the subject of much legislation, and operate as legal bodies within restrictions imposed by numerous statutes; as are Statutory Corporations, such as the nationalized or public industries and undertakings, local councils and police forces – all are legal bodies required to work within the powers and restrictions stated by Acts of Parliament. Government departments and the office of the Crown are also legal bodies capable of suing and being sued.

Liability and insurance

Public liability: goods and products

Designers are legally responsible for defects in the goods and products they have created for manufacture, production and use, not only to their client under the contract for design services, but also to end-users.

Statute and common law together require that any design defects that cause or contribute to an end-user of the goods or products suffering injury, damage or loss, must be remedied by the designer. Remedies might include free replacement or re-design, and payment of compensation.

Goods and products must be 'of merchantable quality' (worth their price) and 'reasonably fit for the purposes required' (function properly and safely, and for a sensible period): in corporate management-speak, 'total quality' means 'zero defects'.

In recent years, the law has plugged many gaps in legal liability that made evasion or avoidance possible (exemption clauses, damage-limitation agreements, notices of disclaimer, registration of guarantees, and so on).

Bluntly, liability cannot be evaded; design defects must be made good. This legal situation merely reflects what for most designers is already best practice, in the home market, and abroad, where similar, sometimes more stringent, provisions exist.

Moreover, certain goods and products are the subject of specific legislation, at home and abroad, setting standards, types and qualities of materials and manufacturing methods that must be used.

Failure to comply can be a criminal offence as well as a civil wrong for which damages or other court orders can be sought.

Designers must equip themselves with sufficient knowledge and understanding of such legislative requirements for goods and products according to the markets in which they will be used.

Insurance against public liability (for all, not just end-users) is available, and is a necessary business expense for designers involved in this field of practice.

Public liability: interiors and structures

Designs for interiors and structures to which people will have access must create a healthy and safe environment free from any hazard.

Overt or latent design defects will render the designer legally responsible, alone or with others. As with goods and products, statute and common law together require 'total quality' and 'zero defects'; it is the same abroad.

In addition, certain places and spaces are the subject of specific legal provisions, at home and abroad, giving minimum standards, types and qualities of materials and methods of construction that must be used to create safe and healthy environments. Criminal liability is frequently imposed, and knowledge of such requirements is therefore vital.

Insurance against liability is often a pre-condition to be met before any contractual deal will be made, and is therefore readily available.

Public liability: advice and assistance

In recent years design consultancies have increasingly offered management and other advisory services to clients, especially to the corporate market, over and above conventional hands-on design work.

In these circumstances, advice and assistance is given by designers to clients at boardroom or executive level in relation to the strategic or managerial development of client companies. Delivery of such services creates legal responsibilities for designers, with which many are not always familiar.

In the UK and abroad, the law protects those who are damaged by acting upon such advice and assistance, by making the givers of negligent advisory services liable to make good any damage such reliance has caused.

Professionally negligent advice (eg that the designer knew or ought to have foreseen might cause trouble, if acted upon) will render the designer liable in law.

Insurance against such negligence claims is available; lawyers and doctors have had it for years and cannot be authorized to practise without it.

Professional indemnity insurance

Whether or not the consultancy is legally accountable to the end user, it will certainly be accountable to its clients for the delivery of services that fall below professional standards that could reasonably be expected of it. Professional indemnity insurance, which can include public liability cover, is available and should always be explored. The Design Business Association runs a Professional Indemnity Insurance scheme for members which is tailored to meet these needs.

Employers' liability

It should be obvious, but is often forgotten, that employers must provide a healthy and safe working environment for their employees and visitors. Insurance is both available and vital.

Legal services

Design consultancies need to know which circumstances make it necessary or desirable to seek professional legal advice and help, and when to do so. Certain matters may be properly and sensibly dealt with in-house; others must go to a lawyer in- or out-house.

Management practices and procedures need to be established and regularly reviewed to ensure that all personnel know the limits of their competence and authority to act, or not to act, and when and how to get the right kind of help. This is easier said than done. The ideal for a large company is to employ an in-house lawyer to deal with corporate business, client contracts and employee training. Most designers use outside lawyers, but only when absolutely necessary (for example, corporate legalities and legal fire-fighting). Accordingly, the vast majority of designers, especially junior employees, have little or no direct access to legal advice and assistance, let alone to the training and coaching needed to dispell their ignorance and equip them with competence.

Ostensible authority

Best practice requires that all personnel know the chain of command authorizing specific levels of competence to commit the business to any legal liability. This means having a working knowledge of the legal concept of 'ostensible authority'.

Design consultancies involving two or more people working together, as a partnership or a limited company, may find themselves embroiled in legal disputes where 'ostensible authority' is a key issue. For example:

- a junior designer/employee finds a new client and commits the consultancy to providing design services, say brand package design, but does this without authority of a line-manager/supervisor. The work is done and paid for, but the client is eventually dissatisfied with the performance of the package in use and seeks compensation for losses caused by an alleged design fault.

Is the consultancy responsible in law, when it did not authorize its junior employee to make any such deal? 'Ostensible authority' could be applied by the law in these circumstances, meaning that the consultancy would be responsible because it had clothed its employee with apparent power to bind the studio. The consultancy had allowed him or her to behave as if authority had been given to make deals, and the client had relied upon that apparent or ostensible authority.

The fact that the employee acted in breach of contract, and perhaps has been sacked accordingly, does not allow the consultancy to evade legal liability.

So, who has what power in making deals?

- ***Companies or partnerships*** are the ultimate authority, but only when making decisions at a legally constituted meeting of the board (of directors) or the partners.
- ***Individual directors or partners*** have no more authority than the most junior employee for the purposes of legally binding the studio. It is vital, therefore, for the board or partners' meeting to specify the extent and degree of each director or partner's executive authority.
- ***Managers***, whether or not they are also directors or partners, likewise need to be given clear written executive authority to act on behalf of the consultancy. The ability to hire and fire, and up to what specified grade of employee, are key matters for clarification.
- ***Employees***, whether or not they are also managers, also need clear written authority to limit, if not to avoid, any damage they might cause owing to their apparent or ostensible authority.

Competencies

When the board of directors or partnership decides to give specific executive authority to individuals working for the consultancy, this should be conveyed in writing through a scheme of delegation, often called line-management, and usually illustrated in a graphic 'organogram'.

In addition to the 'line-management organogram' given to all personnel, best practice requires that everyone is given written terms and conditions of their employment, directorship or partnership, and is instructed or trained to understand where authority exists throughout the whole studio and in relation to themselves (see chapter 14):

- in-house arrangements along these lines should mean that all personnel are informed and understand the nature and extent of their own and others' competence
- ideally, everyone should also be told where, when and how legal advice and help is to be found and taken inside the organization
- out-house legal advice and help from, say, the studio's retained solicitors, will be needed in certain situations, which should be specified to all personnel, even though only a handful of the more senior people will have been authorized by the consultancy's board or partners to engage such out-house legal services.

Best practices

Well-run design practices will have constructed and implemented provisions to cover at least the matters set out above, and probably many more. Doing so will not evade legal liability for any person who exceeds their authority (if any) to commit the consultancy. It will, however, enable such defaulting personnel to be sacked for breach of contract, and to be sued by the studio for reimbursement of any compensation it has been obliged to pay to independent contractors 'ostensibly' hired to provide goods or services.

Summary

It has been the function of this chapter to equip designers at all levels within any consultancy, as well as sole traders, to understand when, where and how to avoid legal pitfalls and to seek professional legal advice. Much of this material is common sense as well as being common law. Many well-run consultancies already know and pursue the practices suggested here as 'best'– for which also read 'absolute minimum'. A valuable rubric for everyone to ponder and to adopt is 'only act within the level of your own competence and authority; this also means knowing and understanding the level of your own incompetence and non-authority'.

' "Write that down", the King said to the jury, and the jury eagerly wrote down all three dates on their slates, and then added them up, and reduced the answer to shillings and pence.'

Lewis Carroll

Alice in Wonderland

4 UNDERSTANDING FINANCE

Bob Willott

Learning to live with the 'number crunchers' has been a long and occasionally painful experience for some design businesses. At one extreme there have been a number of design businesses which have developed an excellent working relationship with their accountants. But at the other extreme there have been a few designers who have seen the management of the business as their exclusive domain and have resented an 'outsider' (meaning someone who is not a designer) participating at that level. They have relegated the financial function to a backroom (or even backyard) role which consists of nothing more than totting up the financial consequences of management decisions, often many months after the event.

The ideal role for a competent financial manager is alongside the rest of the management team: joining in, and committed to, their collective decisions after a full review of the financial implications. Of course, such a person must have the necessary skills and personal qualities (not all accountants would take to the cultural style of a design company) and the role described does not diminish the importance of routine bookkeeping. Indeed it adds to it. Any financial manager who participates in management decisions must also accept the obligation to provide timely, accurate and comprehensible reports not only on the financial consequences of those decisions but also on the financial progress of the business generally.

There are three main roles encapsulated in the title of financial manager: planning, reporting and reacting.

Planning involves the preparation of forecasts, that is trying to put figures on the management's best estimates of future activity. ***Reporting*** involves recording the financial transactions as they take place and providing the management with a prompt and meaningful picture of how the business is progressing. ***Reacting*** means using the financial reports to make constructive management decisions, preferably avoiding the pain of an unpredicted financial crisis.

Financial reports as tools of management

Good financial reports are not just useful, they are absolutely essential. Their purpose should be to highlight key statistics about the performance of the business in a manner which may be digested quickly, enabling recipients to see where action may be necessary and what that action should be. A simple example is the collection of debts. If collection is falling below an acceptable level, this should be immediately evident from a well-designed financial report along with the extent of the deterioration. This will enable the management to see what effect there would be on the bank balance if the deterioration were to be corrected. There are many types of financial report, but for convenience they may be may be divided into three categories: budgeting, profit management and cash management.

- ***Budgeting*** is the prediction of income, expenditure and profit, based on a combination of judgement and all the evidence available. It will never be accurate. It enables the management to assess the range of financial consequences of various budget assumptions. Thus it is a planning exercise.
- ***Profit management*** means the recording and reporting of trading results in a manner which aids decision-making. If profit margins are declining, the financial reports should show why and by how much. As a result decisions can be taken to improve matters (or in exceptional cases to consciously decide not to improve matters). The term applied to reports which help the management of profit is 'management accounts'. They are produced at regular intervals (usually monthly, but with some key data, like new orders, invoices rendered and the cash position, produced weekly).
- ***Cash management*** embraces both monitoring the current position and projecting what may happen in the future. Reports on the cash position usually comprise a statement of the current situation compared to agreed bank facilities, projected progress on debt collection, short-term payment obligations, details of unpresented cheques, and so on. In addition there should be regular forward projections (cash flow forecasts) which attempt to show what will happen to the bank balance over the ensuing 12 months in best- and worst-case scenarios.

All these reports may seem rather daunting, particularly to a small business. Indeed, it would be unrealistic to expect a small business to produce such reports in a comprehensive way. However, it is essential to produce reliable data which are compatible with the scale of the business activity. To produce nothing is to leave financial prosperity entirely to chance, and no-one needs to do that.

Budgetary control

Many a young entrepreneur has dismissed the idea of budgeting as unnecessary bureaucracy. 'I know exactly what my costs are and what I have to sell to cover them', they say. And, if they are right in that assertion, they will not be pleased to be told that they already have a budget, a mental budget. There is nothing mystical about budgeting. It is simply a methodical way of estimating the future costs and revenues of the business. The process enables the management to estimate the break-even position (the amount of sales required to cover costs), as well as to see the impact on profits of various levels of sales income.

In big companies the budgeting exercise tends to be rather extensive, partly because no single person can control the overall financial impact of myriad decisions among a multitude of divisions located all round the world. But that does not mean budgeting needs to be complex in a small business. Nevertheless, it is desirable to commit it to writing and to challenge the assumptions which lie behind it.

One of the biggest benefits of budgeting is in testing the validity of those underlying assumptions. Is it possible to handle a 40-per-cent increase in sales without more people or larger premises? Has proper allowance been made for the interest charges which will arise after buying the new computerized design equipment? What happens if sales fall short of budget by 40 per cent?

Another benefit is in committing people to business objectives. A business performs best when it has a sense of direction. It is not a good idea if the managing director (less still the financial director) creates a budget in isolation from the rest of the senior management and then imposes it on them. It is far better if the senior people put together their initial projections on sales, profit margins, staffing requirements and the like. They will feel more involved and committed to achieving those projections as a result, even if, as so often happens, the managing director has to persuade them either that they can achieve higher sales than they are projecting or that they can manage on less resources. The third major benefit of budgeting is that it provides a yardstick against which subsequent performance may be measured. That is what budgetary control means. Depending on the size of business, the actual financial results (the management accounts) may be compared with the budget at weekly, monthly or quarterly intervals.

Need for an underlying business development strategy

One of the dangers of budgeting is that it can become a routine procedure carried out with little enthusiasm simply to satisfy the accountants. Yet the exercise is meaningless unless the management has developed some clear ideas about where they want to take the business. What competitive pressures will affect the business over the next few years? How will the business preserve or enhance its competitive advantage? The answers to questions like these are crucial to the preparation of a meaningful budget. They will affect projected gross profit margins, geographical coverage, types and numbers of staff, investment costs and many other financial items.

Even the smallest business needs to have a sense of strategic direction and to plan its finances accordingly. So the annual budgeting process provides a useful spur to review those more fundamental issues.

Making budgetary assumptions

Figurework in budgeting is the product of underlying assumptions about market

Preparing a budget: making the underlying assumptions

Assumptions about income

How much will the total market for your services grow/decline in the period?

Will your share of your markets grow/decline?

Will you increase/decrease your prices?

Will you be introducing any new services or expanding into new markets?

Will you be closing down any services or withdrawing from any existing markets?

↓

Assumptions about profit margins

What mark-up do you aim to achieve on external production costs?

Will there be competitive pressure to reduce margins?

Will you be increasing your hourly labour costs? If so, by how much?

What ratio will your charge-out rates bear to labour costs?

↓

Assumptions about costs

Which costs are fixed and which will rise with inflation?

What inflation rate have you assumed?

Which resources will increase/decrease (labour, premises, equipment, borrowings)?

Have you allowed for equipment leasing costs and ad hoc items like lawyers' fees?

↓

Budget preparation

Designco Limited: budget

Year ending 31 December 1992

	Jan £000	Feb £000	March £000	April £000	May £000	June £000	July £000	Aug £000	Sept £000	Oct £000	Nov £000	Dec £000	Total £000
Turnover													
Fees receivable	36	47	56	44	44	44	40	36	36	44	44	44	515
Rechargeable items													
amount invoiced	60	36	40	40	40	40	35	30	40	40	40	40	481
cost of sales	48	29	32	32	32	32	28	24	32	32	32	32	385
Profit margin	12	7	8	8	8	8	7	6	8	8	8	8	96
Gross income	48	54	64	52	52	52	47	42	44	52	52	52	611
Design costs													
salaries etc	15	15	15	15	15	15	17	17	17	17	17	17	192
motor/travel	1	1	1	1	1	1	1	1	1	1	1	1	12
staff expenses	0	1	0	1	0	1	0	1	0	1	0	1	6
equipment hire	1	1	1	1	1	1	1	1	1	1	1	1	12
consumables	0	0	1	0	0	1	0	0	1	0	0	1	4
	17	18	18	18	17	19	19	20	20	20	19	21	226
Gross profit	31	36	46	34	35	33	28	22	24	32	33	31	385
Overheads													
Premises:													
rent and rates	6	6	6	8	8	8	8	8	8	8	8	8	90
light and heat / cleaning / repairs / furniture depn / insurance	1	2	1	1	2	1	1	2	1	1	2	1	16
Marketing													
salaries etc	2	2	2	2	2	2	2	2	2	2	2	2	24
motor/travel / entertaining	1	1	3	1	1	1	1	1	4	1	1	1	17
advertising/PR	1	1	8	2	0	1	0	0	3	1	1	1	19
Administration													
salaries etc	3	3	3	3	3	3	3	3	3	3	3	3	36
motor/travel	1	1	1	1	1	1	1	1	1	1	1	1	12
staff expenses	1	1	1	1	1	1	1	1	1	1	1	1	12
telephone/fax	2	1	2	1	2	1	2	1	2	1	2	1	18
post/messengers	1	1	1	1	1	1	1	1	1	1	1	1	12
other expenses	0	0	1	1	0	1	1	1	0	1	0	1	7
Finance and legal													
interest	1	1	2	1	1	2	1	1	2	1	1	2	16
accounts dept	3	3	3	3	3	3	3	3	3	3	3	3	36
legal fees	0	0	0	0	0	5	0	0	0	0	0	0	5
	23	23	34	26	25	31	25	25	31	25	26	26	320
Other income	0	0	1	0	0	1	0	0	1	0	0	1	4
Profit (loss)	8	13	13	8	10	3	3	(3)	(6)	7	7	6	69

conditions and many other internal and external factors. So how should these assumptions be made? The chart on page 35 highlights some of the key questions to be addressed.

Some people claim that it is impossible to budget for revenue in a design business. The reason given is that design work is project based whereas in the advertising business, for example, clients have annual spending commitments. However, the doubters should be aware that there are many other businesses in a similar position to design (eg solicitors, shopkeepers and sheep-farmers) which manage to budget for revenue efficiently.

Those who are good at budgeting for revenue recognize that they must consider market forces such as the economy and the strength of competition, and not simply assess their likely cost base and then arrive at the revenue projection by multiplying costs by a mark-up. That is not budgeting. It is the art of self-delusion.

An almost inevitable consequence of proper budgeting is that, during the process, revenue projections will at first be out of line with cost projections.

The gap between revenue and costs will be either unusually large or unnervingly small. At this stage it is necessary to review both parts of the budget to see whether there is undue optimism or undue pessimism. If the profit gap is very small and the projections are deemed realistic, some serious decisions will be necessary. Clearly the cost base will need to be pruned unless additional revenue can be obtained from, for example, increasing charge-out rates or increasing margins on rechargeable production costs.

Example of a budget

The budget on page 36 shows what a fairly simple budget might look like. The amount of detail will depend on the size of the business. A large business will have budgets for each operating division, showing the profit contribution from each. These divisional budgets are then brought together to produce (hopefully) an aggregate profit contribution from which is deducted any central overhead costs which the management believes should not be allocated to any specific operating division. However, a smaller business will need just one page for the entire exercise.

Overhead costs are those items which are not incurred directly in the production or supply of the services or products of the business. Depending on the nature of the business, they would probably include administrative accommodation, professional fees and receptionists' salaries, for example. In preparing a budget, most overhead items are spread evenly across the 12 months irrespective of when the bills are received or paid. In this way costs are more evenly matched with income.

Of course, those costs which vary directly with income will be budgeted on that basis. For example, the cost of items rechargeable to clients (rechargeable items) appears as a constant percentage of the amount invoiced to those clients on the assumption that the business policy is to mark up such expenditure.

In managing the finances of a design business it is particularly important to separate the design work itself from any goods procured for the client from outside the business such as printed literature, and shop-fittings. This enables the management to budget for the productivity of designers as well as for the margin to be earned on rechargeable items bought from outside.

The format used in the example is consistent with that of the periodic management accounts (page 38). This makes it easy to compare actual performance with budget.

Designco Limited: management accounts

Month: September 1992

	This month			Cumulative		
	Actual £000	Budget £000	Variance £000	Actual £000	Budget £000	Variance £000
Turnover						
Fees receivable	41	36	5	412	383	29
Rechargeable items						
amount invoiced	26	40	(14)	354	361	(7)
cost of sales	21	32	(11)	292	289	3
profit margin	5	8	(3)	62	72	(10)
Gross income	46	44	2	474	455	19
Design costs						
salaries etc	15	17	(2)	159	141	18
motor/travel	2	1	1	10	9	1
staff expenses	0	0	0	6	4	2
equipment hire	2	1	1	13	9	4
consumables	0	1	(1)	4	3	1
	19	20	(1)	192	166	26
Gross profit	27	24	3	282	289	(7)
Overheads						
Premises						
rent and rates	8	8	0	67	66	1
light and heat, cleaning, repairs, furniture depn, insurance	2	1	1	13	12	1
Marketing						
salaries etc	2	2	0	17	18	(1)
motor/travel, entertaining	5	4	1	16	14	2
advertising/PR	6	3	3	20	16	4
Administration						
salaries etc	2	3	(1)	28	27	1
motor/travel	1	1	0	8	9	(1)
staff expenses	1	1	0	10	9	1
telephone/fax	3	2	1	16	14	2
post/messengers	1	1	0	8	9	(1)
other expenses	1	0	1	7	5	2
Finance and legal						
interest	1	2	(1)	9	12	(3)
accounts dept	3	3	0	28	27	1
legal fees	2	0	2	9	5	4
	38	31	7	256	243	13
Other income	1	1	0	3	3	0
Profit (loss)	(10)	(6)	(4)	29	49	(20)

Managing profits

Assuming the budget has been carefully prepared, any significant deviation (variance) in the actual performance of the business merits examination. And the sooner the management accounts reveal the deviation, the sooner corrective action may be taken. The objective should be to 'beat the budget' in all key areas: better sales revenue, better profit margins and lower costs. In practice, it is rare to win on every count.

Often businesses start spending in response to improved sales without adequate regard to the ratio (arithmetical proportion) of that spending to the improvement in sales. As a result profit margins fall. The outcome can be a nasty shock if and when sales slide back down again.

Variances from budget and how to respond to them

It is a good idea to feed details of variances in actual performance from budget back to the manager with the day-to-day ability to act upon them. The reason(s) for the variance should be explored (it might be just an accounting error) and appropriate action taken or recommended. A manager who has been actively involved in preparing the budget will usually be keen to achieve or beat it.

Page 38 shows a typical format for management accounts. It shows:

- results for the latest month (September) compared with the budget for that month
- the variances (adverse or favourable) for each item (see the third column); where actual performance is lower than budget, the variance appears in brackets
- the cumulative results for the year to date alongside the results for the latest month; again these are compared with the cumulative budget and variances are identified.

Designco's results show a shortfall against budget for the latest month and cumulatively. The shortfall is modest in the latest month and arises mainly from:

- an overspend on overheads (£38,000 against a budget of £31,000)
- a shortfall in profit on rechargeables (£5,000 against a budget of £8,000) the effects of which were reduced by a £5,000 surplus of fee income over budget.

The cumulative picture is a little more worrying:

- despite fee income exceeding budget by a significant amount, in percentage terms design costs have raced ahead even faster
- the cumulative profit on rechargeables is running behind budget
- the overspend on overheads appears to have been building up over a period.

With only three months left before the year end (and even less by the time the management accounts would have been produced), it may be difficult to recover all the shortfall in profits during the remainder of the period.

Key performance measures

To make the most of the management accounts, a limited number of key performance measures should be highlighted.

Designco Limited: performance indicators

	This period		Year to date		Previous
	Actual	Budget	Actual	Budget	year
Gross income: design costs	2.4	2.2	2.5	2.7	2.6
Profit margin on rechargeables %	19.2	20.0	17.5	19.9	20.0
Gross profit/gross income %	58.7	54.5	59.5	63.5	61.5
Overheads/gross income %	82.6	70.5	54.0	53.4	57.2
Operating profit/gross income %	(21.7)	(13.6)	6.1	10.8	10.9
Trade debtors/ turnover (days)	67	55	61	55	58
Excess trade debtors over budget (£000)	32	-	32	-	13
Per designer per annum:					
gross income	27.6	26.4	31.6	30.3	30.2
design costs	11.4	12.0	12.8	11.1	11.6
overheads	22.8	18.6	17.1	16.2	17.2

Performance measures may include:
- the ratio of gross income (fees plus margin on rechargeables) to design costs or design labour
- the profit margin on rechargeables
- overheads as a percentage of design costs or design labour
- overheads as a percentage of gross income
- trade debtors as a percentage of turnover
- gross income, design costs and overheads per designer.

All this information may be presented in the form of a summary attached to the management accounts, including comparisons with the previous year and with the budget where appropriate (see page 39). If the management accounts show a deterioration in any of the key statistics, this should prompt remedial action.

Maximizing profit margins and controlling costs

It is hard to maintain profit margins at the best of times and particularly so during a recession. But there are several opportunities to do so. For example:
- expenditure on rechargeables may not have been adequately recorded and recovered because the job bag system is inefficient and the client is not invoiced for every cost incurred. Sometimes errors of this sort can be recovered later, but this depends on the client's goodwill and is not good practice.
- margins may be improved by a deliberate increase in the mark-up applied (whether across the board or on certain projects). This is not easy to achieve, especially if business is less than bouyant.
- the margin on design work may be improved either by increasing charge-out rates, if they have fallen behind market rates, or by cutting back non-chargeable hours (even to the extent of making people redundant if the situation looks like remaining depressed for some time).

Controlling costs involves enforcing appropriate approval procedures and even having the occasional witch-hunt. Sometimes unpopular steps, such as closing or restricting the use of a cab account, are necessary simply to bring home to everyone the importance of watching costs. However, savings need to be achieved in a way which does not damage the routine servicing of clients.

On the cost front it is also useful to remind colleagues that for every £1 of extra cost, the business needs to earn an extra £3 in fees or profit on rechargeables if margins are to be maintained. Where that level of extra revenue is not confidently expected in the very near future, the expenditure must be questionable.

Focusing on material issues

One of the dangers of generating too much management information is that managers delve into every minor variance. The art is to focus on the key issues and to take remedial action quickly and decisively. If the rechargeable items tend to be a small part of the turnover, any variation in mark-up is unlikely to make much impact on the bottom line. Similarly there is little point in pressing the bank manager to prune back the interest rate if the level of borrowings is so low as to render any saving of little consequence. On the other hand, property costs and labour costs will usually be among the largest items. In boom times there is always a temptation to take on more space in anticipation of further growth. But the impact on profits and cash can be substantial, not to mention the disruption which arises. In a recession, it may be very difficult to dispose of excess accommodation.

Designco Limited: cash flow forecast													
Year ending 31 December 1992													
	Jan	Feb	March	April	May	June	July	Aug	Sept	Oct	Nov	Dec	Total
Receipts	£000	£000	£000	£000	£000	£000	£000	£000	£000	£000	£000	£000	£000
Fees /rechargeables	90	91	92	85	90	87	85	82	76	71	75	81	1,005
Other income	0	0	1	0	0	1	0	0	1	0	0	1	4
VAT on above	16	16	16	15	16	15	14	15	13	13	13	14	176
Total receipts	106	107	109	100	106	103	99	97	90	84	88	96	1,185
Payments													
Cost of sales	41	48	29	32	32	32	32	28	24	32	32	32	394
Consumables	1	0	0	1	0	0	1	0	0	1	0	0	4
Other costs:													
salaries	12	12	12	12	12	12	13	13	13	13	13	13	150
PAYE/NIC	8	8	8	8	8	8	8	9	9	9	9	9	101
expenses/motor	4	5	6	5	4	5	4	5	7	5	4	5	59
telephone/fax			4			5			4			5	18
post/messengers	1	1	1	1	1	1	1	1	1	1	1	1	12
rent/rates			20	6		20			20	6		20	92
light/heat	3			4			2			2			11
cleaning/repairs			1						1				2
insurance					3								3
eqpt hire etc	1	1	1	1	1	1	1	1	1	1	1	1	12
advtg/PR	1	1	1	8	2	0	1	0	0	3	1	1	19
other expenses	1	0	0	1	1	0	1	1	1	0	1	0	7
interest			4			4			4			4	16
accounts dept	2	2	2	2	10	2	2	2	2	2	2	2	32
legal	0	0	0	0	0	0	0	5	0	0	0	0	5
Capital payments:													
equipment cost	0	0	0	0	0	0	0	40	0	0	0	0	40
corporation tax	36												36
	111	78	89	81	74	90	66	105	87	75	64	93	1,013
VAT													
on inputs	10	10	8	10	9	8	8	15	7	8	7	8	108
remittances	21	0	0	20	0	0	19	0	0	8	0	0	68
Total payments	142	88	97	111	83	98	93	120	94	91	71	101	1,189
Net cash flow	-36	19	12	-11	23	5	6	-23	-4	-7	17	-5	-4
Cash b/fwd	43	7	26	38	27	50	55	61	38	34	27	44	43
Cash c/fwd	7	26	38	27	50	55	61	38	34	27	44	39	39

Inevitably staff levels and remuneration must always be a candidate for management attention as, without constant monitoring, costs can move out of line with revenue. Sadly, they are one of the few areas where meaningful savings can be achieved, however painful that process might be.

Managing cash

The natural tendency in budgeting is to produce a projection of what cash will be received and what payments will be made. That is not a budget. A budget should show the revenue and costs arising from running the business at the time they arise, unaffected by when cash is actually collected or paid out. The budget should focus only on items which will contribute to the operating profit or loss and should exclude, for example, capital expenditure (which in a larger business may be projected in a separate 'capital expenditure budget').

A projection of cash movements (their timing and amounts) is normally prepared separately, combining transactions relating to the profit and loss account with those which are capital in nature (that is, everything else). Such a projection is commonly called a cash flow forecast. This forecast may be needed to support an application for bank facilities and is a useful management discipline in any event (see page 41).

Benefits of cash forecasting

The most obvious benefit of cash forecasting is that it keeps a business ahead of the game in its banking relationships. First, it should show the likely borrowing requirement and it will be possible to see whether the ratio of projected borrowings to the funds invested in the business by its owners will be acceptable (banks do not like borrowings to exceed the owners' stake.)

A cash forecast will also help the management assess whether there is likely to be sufficient debtor cover as security for the bank. Banks like the amount they lend to a customer to be covered at least twice by the amount of debts due to that customer by its clients. Sometimes a bank may be looking for debtor cover of three times the amount lent to the customer if they regard the situation as risky. To some extent that will depend on the quality of the clients as a credit risk and whether they are credit insured (insurance companies will often insure a business against the risk of a bad debt, but clients in a weak financial condition may not be insurable).

If the borrowing requirement is likely to exceed normal banking tolerance levels, action can be taken well in advance. If the manager seems reluctant to provide the necessary facilities, and the requirement is not excessive, there will be time to shop around at other banks. Similarly, if the bank is willing to help but wants an unusually high rate of interest or excessive security, there is time to look at alternatives before the peak borrowing requirement is reached.

It is invariably a mistake not to keep a bank manager 'on side'. Bank managers are in business to lend money, albeit they may sometimes seem to want the minimum risk for the maximum reward. Good managers will wish to share in the management's thinking and to help them succeed. But no bank manager likes a surprise, especially a nasty one. The banking system is managed hierarchically and facilities may have to be approved by someone beyond the manager, possibly at another location. Often the manager will discuss alternative ways of packaging the funding required, partly to take account of what his or her superiors are likely to approve. So it is worth listening to what bank managers have to say and recognizing that they have a job to do.

If the worst comes to the worst, and the bank will not provide all the finance required, there should still be time to look at alternatives, ranging from simply re-timing certain capital expenditure plans, improving debt collection, arranging for a further injection of capital by the owners (if they have funds available), or looking for long-term capital, possibly in the form of equity (meaning funds provided in return for a share in the ownership of the business) from an outside party.

Cash forecasting will help the managers of the business to plan the timing of key transactions. An obvious example would be the withdrawal of profits. Just because the business has made a healthy sum, it is not necessarily wise to draw it all out. Expanding businesses tend to swallow up cash in funding the work in progress (paying staff before the cost can be invoiced to clients) and debtors (whilst waiting for the invoices to be paid). A cash flow forecast should highlight whether, when and how much of the profits may be distributed to the owners of the business.

A further benefit is the ability to anticipate the size of any cash surplus which may be generated and to make the best use of it. Often a better return can be earned than simple deposit interest, although businesses need to be careful about where they invest surplus funds, as the 1991 BCCI affair illustrated. Most banks are able to provide opportunities for maximizing the return on deposits (for example, by using the overnight money markets) but it may be necessary to ask before the service is offered.

How cash forecasting is done

To prepare a cash flow forecast, it is necessary to rework the information included in the budget (see page 36) to reflect the timing of each transaction. For example, an assumption will need to be made about how swiftly clients pay their bills. Perhaps it would be reasonable to assume that 10 per cent of clients will pay within 30 days, 50 per cent might pay within 60 days, 35 per cent within 90 days and the remaining 5 per cent might pay within 120 days. In cash forecasting it is important to err on the side of caution. There is no sense in preparing a projection which will not be achieved, because this will simply undermine the bank's confidence.

Assumptions will also have to be made about the timing of payments to creditors, including the Inland Revenue and HM Customs & Excise (who collect VAT). It is not just the monthly PAYE payments that have to be considered. There is also the timing of annual corporation tax liability (assuming there is one). Then there is the whole area of capital payments, from hire purchase or leasing instalments to one-off payments for the purchase of equipment. Costs of moving premises (including agents' and lawyers' costs) may also have to be contemplated.

All the underlying assumptions should be written down and their appropriateness discussed. Then the assumptions should be applied in compiling the forecast itself. Once completed, it is wise to play with the numbers to see what impact there would be if, for example, debts could be collected quicker or sales fell short of expectations. Such an exercise is called a sensitivity analysis because it examines the sensitivity of the forecast to particular changes in underlying assumptions. Some changes may have suprisingly little impact on borrowing requirements; others may create a shock.

Some key ratios

As with the management accounts there are some key ratios which need to be watched in monitoring the cash position. Most are based on balance sheet information. If possible a balance sheet should be produced as part of the budgeting and cash forecasting process, but this will depend on the size of the business and its accounting capability. Sometimes financial advisers will have standard computer packages which can handle the entire budgeting and cash forecasting exercise for those clients who do not have the capability themselves.

The guidelines to good cash management include ensuring that:

- there is a surplus of net current assets, (eg debtors, work in progress and cash), less current liabilities like PAYE, other ordinary creditors and bank overdrafts
- the debt:equity ratio remains lower than 1:1; 'debt' means external borrowings and 'equity' means the owners' capital invested in the business (whether in shares, retained profits or loans)
- there is adequate debtor cover to secure bank borrowings; 'debtor cover' means the ratio of trade debtors (money due from clients) to bank borrowings
- the amount of cash tied up in trade debtors and work in progress is reduced to the minimum by good credit control procedures.

Ian Whadcock/Sharp Practice

'It is impossible to teach a person anything: you can only help them to find it within themselves.'

Galileo

5 USING THE SPOKEN WORD
Shan Preddy

This chapter deals with communication in all of its spoken forms, specifically the issues involved in good spoken communication and practical advice on giving presentations, handling important phone calls, and running successful meetings. Chapter 9 develops this subject further.

Spoken communication

Communication does not start and end with one person. The word 'communication' comes from the Latin *communicare*, meaning to share. Communication involves, at its most basic level, a message sent by one party, and received by another.

Historically, and across all cultures, the most common forms of communication are visual (drawn or illustrated materials, and gestures) and verbal (written and spoken).

Spoken communication takes place every day, both socially and commercially, with friends and with colleagues. Why then, if it is so commonplace, is it necessary to think seriously about spoken communication in business? Why are skills in spoken communication important for a design company, which is, after all, particularly expert in visual communication?

Design companies operate in a fast-moving business environment, with much of the communication both inside and outside of the company based on the spoken word, whether in presentations and meetings or in telephone conversations. Design professionals are increasingly assessed by their clients not only on their creative and commercial ability, but also on their skill in communicating that ability through the verbal medium. A design company whose staff can provide effective creative solutions, but cannot explain those solutions in a clear, persuasive and articulate manner, will not succeed. A brilliant presentation can, at least temporarily, mask poor design work; a sub-standard presentation will fail to support otherwise excellent work.

Spoken versus written communication

Spoken and written forms of communication each have their advantages; choosing which to use depends on the situation and on the objectives.

Advantages of spoken communication

- the spoken word is often more persuasive than a written letter or document, which is unable to assess the recipient's reactions immediately, or to give an instant response to any objections raised
- both forms of communication can be tailored to the individual recipient; however, only spoken communication can make alterations half-way through

- spoken communication is generally two-way. Exceptions are: appearances on television and radio; video, film and audio tape; public address systems; messages on answerphones. Two-way communication adds clarity to the message, and makes it possible to gain useful and immediate aural or visual feedback.
- the meaning of spoken communication is clear; written communication can be open to misinterpretation. Content is given meaning by tone of voice, emphasis on certain words, and, in face-to-face communication, gestures.

Disadvantages of spoken communication

- it is inadequate for imparting information, such as technical plans, detailed specifications, timing schedules or cost estimates; people have a limited ability to retain complex spoken information
- it is transient, and cannot, unless audio recorded, be kept for future reference. Verbal agreements should be confirmed in writing and distributed to all interested parties; examples might include letters, contact or call reports, and documents. (See chapters 3 and 15 on legal issues, and chapter 6 on the written word.)

The skills of a good spoken communicator

What makes a good spoken communicator? Skill in speech patterns, appearance and body language all help to retain a listener's interest and attention. (For more guidance in this area, see *Further reading*.) Significant improvements can be made at all levels of experience by investing in personal presentation skills training.

Clarity of thought and the ability to choose the appropriate words and style for the audience and occasion are also important. A description of an aspect of the design process would be given differently to an experienced designer and an inexperienced design buyer, because of their respective levels of knowledge.

Jargon, in all of its forms, whether design, marketing or financial, is only helpful as a form of shorthand if the recipient speaks that jargon fluently. If the recipient's knowledge of that jargon is poor, misunderstandings and mistakes will happen. Unfortunately, it is often difficult to admit in a business situation that something has not been understood. The speaker is responsible for ensuring that the message has been not only received, but also understood. A good spoken communicator will constantly monitor the feedback from the other party, and adapt and adjust the message accordingly.

The single most important attribute of a good communicator is the ability to think about the recipient of the communication, and to prepare and act accordingly. This is the key to the art of persuasion.

Any spoken communication should aim for the accurate delivery of a message, expressed in words, sent by one party, and received by another. However, receipt of the message is not enough. The sender needs to consider the effect of that message on the recipient. In other words, what should the recipient feel, think, believe, understand or do as result of receiving the communication? One of the greatest faults in communication is the failure to establish the desired outcome.

Business presentations and telephone calls

The suggestions which follow apply primarily to face-to-face presentations, whether of the company and its abilities, or of creative proposals and ongoing work. However, the recommendations apply equally to telephone calls. (See Collier Cool 1989 for specific advice on telephone skills.) Recommendations on the content of sales meetings and telephone calls are given in chapter 10.

It is essential to prepare for all presentations and important phone calls, whether they are to clients, the boss, colleagues, juniors or suppliers. The quality of the preparation will affect the final result. 'Fail to prepare, and you prepare to fail' applies to presentations and phone calls as well as other aspects of business life.

The following checklist can be used to help the process of preparation. Once the system is thoroughly familiar, preparing for any presentation, however unexpected, will be swift and efficient.

1 Objectives

To set the objective of a presentation or telephone call, establish the desired outcome:

- what exactly is to be communicated? Why?
- what should the other party think, feel, believe, understand or do as a result of the presentation?

2 Audience

The most important participant in any presentation is not the presenter, but the recipient. The following points are useful in assessing an audience:

- is the audience senior in status? All clients, however junior in years or experience, are senior. Is the audience composed of peer group colleagues, or juniors? The content and style of the presentation may need to be adapted.
- there is an advantage in presenting to colleagues or people who are well known; it is easier to prepare the content and to assess audience. However, there are also disadvantages; the audience may have preconceived ideas about both the subject and the presenter, and the presenter might assume a greater knowledge about the subject by the audience than is, in fact, the case.
- the key people in the audience need to be identified, so that the presentation can be developed at the correct level for them. It is rare that every member of a meeting is a key decision-maker.
- a view on the level of the audience's existing knowledge on the subject helps. What are they likely to know? More importantly, what do they need to know?
- it is important to establish how many people will be present. This will affect both the style of the presentation, and the choice of visual aids.

3 Tone and style

Presentations range widely in tone and style; the choice should be appropriate for the audience, and for the subject matter:

- should it be very formal, conveying detached professionalism?
- should it be warm and friendly, thereby creating and building on rapport with the audience?

In the design business, a formally structured presentation, given in an informal style, can be very successful.

4 Content

The content of any presentation should be carefully considered, and structured in the most appropriate way. All or some of these factors will influence content:

- the balance of information and persuasion needs to be determined; almost all presentations are a combination of the two
- the content should be structured in a logical order, which will allow the audience to follow the argument

- there should always be a beginning, a middle and an end to the presentation. In other words: an introduction; an exposition of the ideas or argument; and a summary and conclusion. 'Tell them what you are going to tell them. Tell them. Then tell them what you have told them.'
- it is dangerous to put the most important point right at the beginning; audiences need a certain amount of warm-up time to concentrate on what you are saying. However, too long a build-up may irritate those listening.
- regular 'signposting' helps. Audiences find it easier to concentrate if reminded where they are in the presentation. An example might be 'Well, that completes our view on the strategy. I would now like to move on to our recommendations for a creative solution'.
- supporting evidence, such as quoted figures, should be to hand
- the argument should be clearly expressed, with the reasoning behind the proposal or recommendation explained. If certain routes have been rejected, it is worth mentioning why; audiences are not mind readers.
- it is sensible to imagine what objections the audience may raise. Rational and logical answers can then be prepared, and either dealt with during the presentation, or reserved for questions.

5 Memory aids

Experienced presenters will always use some kind of memory aid:

- brief bullet-point notes can be prepared, preferably on one side of paper. If the presentation is to be given from a standing position, the notes need to be clear enough to read from a distance. Notes held in the hand always distract from the content; if this cannot be avoided, use index cards rather than paper.
- a presentation should not be read in full except at a conference, when this is not only acceptable, but advisable.

6 Visual aids

Speech generally runs at about 130-150 words per minute; an audience can process information at about three times that speed. As a result, concentration starts to lapse after about ten minutes, however interesting the subject matter. This can be countered by the use of visual aids. Aids, such as presentation boards, 35mm slides, 3D models, and handouts, can help to vary a presentation, to explain points, and to keep an audience's interest. The choice and handling of visual aids need careful consideration; if in doubt on this issue, it is worth obtaining advice from colleagues, or investing in professional training.

The presenter is an important visual aid. Appearance and behaviour should be used to good effect (see Davies 1991).

7 Edit

The content should be structured for maximum effect; now is the time to test it.

- is it concise?
- will it be clear to the audience in question?
- are the facts accurate?
- does the argument work towards the established objective? Irrelevancies should be avoided: they will waste time and confuse the issue.

8 Rehearse

For a major presentation, time should be scheduled for a rehearsal and to make any necessary alterations, particularly if more than one person is presenting:

- if the argument is difficult, it can be helpful to test it out on someone
- if the presentation is formal, a full rehearsal is advisable, with, if possible, a dummy audience. Otherwise, practising in front of a mirror can reap benefits. The presentation can also be audio or video recorded.
- if allocated a certain timespan, the presentation should be prepared and rehearsed accordingly, using a stop watch. If the presentation is too short, it will not make the best use of the time available. If it is too long, it will create difficulties for the meeting, and could alienate the audience.
- if the room in which the presentation will be given is known, or seen beforehand, it helps to relax the presenter
- the decision on whether to sit or stand largely depends on the size of the audience. However, persuasion is easier from a seated position, while information is more authoritative when given by a standing presenter. It is possible to stand at the beginning of the presentation, when a firm point of view is being delivered, and then sit to persuade the audience that the point of view is right.

While the presentation is in progress, it is useful to monitor audience response. Adjustments may need to be made to timing, content or style. Eye contact should be maintained; people who will not meet the eye are seen as untrustworthy. In addition, audience reaction can be observed. It is easy to tell when people are agreeing, but apparent frowns should be treated carefully. People frown when they are concentrating, or when they cannot hear clearly, as well as when they disagree with what is said. Lastly, people need time to accept new ideas of any kind. Presenters are often so familiar with their material, or so convinced by their own arguments, that they tend to bulldoze the audience through a decision process. Resentment and objections inevitably arise.

Remedies for nerves

Some people feel nervous all of the time. All people feel nervous some of the time. Nerves are normal, universal and not usually noticeable by others. Nerves are also beneficial; the adrenalin sharpens up thought processes and improves memory.

The remedy is to be well prepared. However, if attacked by nerves, breathe deeply and steadily, relax shoulder and facial muscles, take a sip of water, and smile. People will automatically smile back, which helps to create confidence.

Meetings

Any gathering with two or more people and a shared purpose is a meeting. At best, meetings are productive and useful. At worst, they are unnecessary and a waste of time. Many executives resent the amount of time they seem to spend in meetings, claiming that it stops them getting any real work done. When thinking of arranging a meeting, the first question to ask is whether a meeting is strictly necessary. Could the information be given in a note, a letter or memo, or by fax? Would a telephone call be better? Could a notice be placed on a board? Would an impromptu chat work? Could the decision be taken unilaterally?

If a meeting is necessary, the following six basic questions will provide a useful guide to planning for success.

1 Why?

The objective and purpose of the meeting should be determined. Without clear objectives, communicated to all participants, no meeting will be efficient. If there is more than one objective, they should be prioritized. The leader or chairperson should state the objective(s) right at the beginning of the meeting.

2 What?

This involves the items to be discussed, or the content. An agenda, whether formally typed or simply jotted down as a memory aid, is essential. If the meeting is to cover a number of items which participants should consider beforehand, it is helpful to circulate the agenda in advance.

The term 'hidden agenda' really refers to hidden objectives. A meeting which is supposedly about getting work approved by a marketing director may also be an opportunity for a brand manager to impress the boss. A design company may call a meeting to present a stage of work in progress, and have as a secondary objective the aim of picking up a further piece of work that they have heard rumours about. Needless to say, items such as these do not appear on the official agenda.

At the meeting, the leader should introduce the agenda, ask the participants if anything needs to be added, and make sure everyone is happy with the timescale of the meeting. The leader should then take part in the discussion, move the discussion on to the next item on the agenda when necessary while keeping an eye on the time, and summarize at the end any decisions that have been taken, and what the next action will be. If each meeting ends with a summary of the positive aspects, it will increase the motivation of those present to carry out agreed actions.

Finally, the results of the meeting should be confirmed by letter, memo, minutes or a short contact/call report, particularly if there may be some debate later about the decisions. The written report will also ensure that people who were not in the meeting are aware of action that needs to be taken.

3 Who?

The essential participants should be agreed; who needs to be at the meeting? Reasons for inclusion might be:

- information to impart
- useful expertise or knowledge about the subject under discussion
- executive responsibility for the subject under discussion
- responsibility for approval
- ability to give support
- need to be informed of the thinking (and it is quicker to do it now rather than report back later)
- likelihood of being upset if not included; a poor but very real reason.

4 When?

The date, time of day, and the length of the meeting need to be set. Apart from ensuring that the key participants are free, and informing all participants when the meeting will take place, the implications of the time chosen should be considered. A meeting late in the afternoon immediately before a public holiday break is not conducive to concentration. A meeting early in the morning following a holiday can be equally non-productive. If a key participant is irritable in the mornings, or sleepy after lunch, those times should be avoided. If people have to travel any distance to the meeting, their needs will have to be considered.

5 Where?

The location and choice of room will affect not only the style but also the success of a meeting. If the available facilities are not suitable, a room should be hired. The room and the environment should be appropriate to the meeting. Two people in a large room can feel almost as uncomfortable as too many people in a small room. Clients who are accustomed to very formal meeting rooms in their company can take a little time to adjust to a different environment in a design company. If confidential matters are to be discussed, or if disciplinary action is necessary, a private room is essential.

Of course, some of the time, there will be no choice. Successful meetings have been held around a desk in an open plan office, on trains, in airports and under many other difficult conditions. On the whole, such meetings are short and to the point.

6 How?

The chosen physical arrangements will influence the outcome. Items for consideration might include:

- setting the seating arrangements
- making sure any necessary equipment is present, and in working order
- lighting, and temperature of the room
- catering requirements
- provision of sufficient numbers of documents, handouts, glasses, coffee cups, chairs, notepads, pencils etc.
- information on how to get to the venue, and parking arrangements
- making sure visitors are expected, and greeted by name when they arrive
- checking that the room is ready, old coffee cups and cigarette ends cleared away.

All of these are things which will only be noticed if they go wrong; unfortunately, praise in this area is rare.

Conclusion

A good communicator is made, not born. All skilled communicators are either experienced, or have been well trained, or both. A certain amount of natural ability helps, of course, but it is not enough on its own. The required skills can be learned, and, once learned, will never be forgotten. However, like all skills, they must be used. Flaubert once said of writing prose that, like hair, it should be combed until it shines. Only frequent practice will keep spoken communication skills shining.

Further reading

Bernstein, David and Audley, Rex (1988) *Put it Together, Put it Across.* London: Cassell.

Berry, Cicely (1975) *Your Voice and How to Use it Successfully.* Bromley: Harrap.

Collier Cool, Lisa (1988) *Phone Power.* London: Robert Hale.

Davies, Philippa (1991) *Your Total Image: How to Communicate Success.* London: Piatkus.

Haynes, Marion (1988) *Effective Meetings Skills.* London: Kogan Page.

Janner, Greville (1989) *Janner on Presentation.* London: Business Books.

Peel, Malcolm (1988) *How to Make Meetings Work.* London: Kogan Page.

Christina Brimage/Sharp Practice

'If language is not correct, then what is said is not what is meant, and what ought to be done remains undone.'

Confucius

6 USING THE WRITTEN WORD

Liz Lydiate

The ability to express thoughts and communicate information in clear written English (or another language) is central to the work of a design consultancy.

Clients often feel out of their depth or unqualified to make aesthetic judgements, but they will be influenced by the quality of written work and the quality of service coming from the consultancy. Although not always justified, most people will link sloppy work in one area with the likelihood of work being equally sub-standard in others. Poor written communication can weaken a client's confidence in the quality of the creative work.

It is often necessary to discuss and explain creative proposals, particularly the more innovative ones, in order to gain client understanding and acceptance. The ability to express complex ideas clearly in writing can be an aid to persuading clients of the value of a particular solution and building their commitment to the creative process.

Design projects involve many complex decisions and transactions, over an extended period. Written communication is necessary to convey proposals clearly and accurately, and to record agreements reached at every stage. 'Get it in writing' is a theme running throughout this book.

Developing written communication skills

Written communication involves making good and accurate use of words, and this can be improved/extended by working on vocabulary and spelling, and giving attention to the way in which words are put together and used. This includes grammar, the rules by which language is constructed, and style, the overall way in which language is applied in order to transfer ideas and information from one person to another.

Everyone learns grammar at school, and it perhaps seems strange to suggest that working professionals go back to such a basic area of study. Because the suggestion is made specifically in relation to working effectively in design consultancy, it has a clear objective, and the learning is directed towards the achievement of known goals, or to offset particular difficulties.

The first step is to decide whether or not to do it, and if 'yes' to make several easy moves towards translating intention into reality.

Laying the foundation: the equipment

It is satisfying to be able to go out and buy new kit in order to learn a new skill; the basic kit for skill in writing includes the following:

- in order to use language effectively in a design consultancy, it is helpful to have a good dictionary and a thesaurus available in the studio at all times; these should be personal rather than communal, and kept in each person's own workspace

- a good dictionary is a long one, such as *The Shorter Oxford English Dictionary* (a huge book) or the more compact *Cassell's.* The advantage of a big dictionary is not only that it has more words, but that it gives a huge amount of information as well, and can be a real source of ideas and understanding. Just owning a dictionary is not enough; it has to be used, and it is helpful if design consultancies can develop a culture where being seen to both own and use a dictionary is regarded as both normal and 'a good thing'.
- the thesaurus offers a range of alternatives to any given word. It is very valuable in breaking deadlocks, where a particular word is being used too frequently, and also provides new ideas when exploring creative concepts. Using a thesaurus is also helpful in copywriting, particularly if a longer or shorter word is needed for aesthetic reasons. People who can write well to a specified length or in a given number of words are greatly loved in design companies.
- designers need to read. It is important for staff at all levels to read a daily newspaper and key trade magazines on an ongoing basis, and to regard this as an essential part of their work. Reading the business pages in daily and Sunday newspapers provides an insight into the world of clients, and a wide knowledge of current affairs is important to producing intelligent relevant design.
- newspapers and magazines publish the work of many highly skilled creative writers, and consciously picking out and analysing good written work is a valuable learning process: 'I enjoyed reading this – why?' 'This information made me stop and think – how did the writer do it?' 'Why is this article so effective?'
- besides work-based reading, there is also private reading. The choice of books for personal use can make an important contribution to developing written work.
- think also about access to library facilities, in-house and externally (see chapter 17)
- many people have access to personal computers with spell-check capabilities. This is a great blessing, but shouldn't be regarded as a substitute for improving spelling and grammar; the machine is only as good as the person using it. Try to feed the pc with good, creative material, and take advantage of its ability to do the donkey work in eliminating slips.

Laying the foundation: the time

Improving written communication doesn't require a lot of time; instead, it takes commitment over an extended period, with small amounts of conscious effort from time to time. This chapter suggests various techniques towards improving the use of written language; it is a good idea to tackle one issue at a time, carry on until it is second nature, and then move on to another area. This ongoing effort is relatively painless, and the visible achievement as the initiative progresses is a valuable stimulus.

Spelling

The DBA Professional Practice Course Stage One has become notorious in the industry for asking delegates to take part in a spelling test. The aim of this is not to humiliate participants, but to make a start towards establishing the scale of the problem privately for each individual. Very few people don't have a problem of some kind with spelling, and, once this is recognized, there are various steps which can be taken to achieve improvement.

The first step is to identify words which are consistently misspelled. Everyone has a group of words which gives them problems; pin these words down and then make a deliberate effort to learn the correct spelling:

- study the word carefully, pronounce it, break it down into syllables
- try to think of a useful mnemonic or trick to overcome the spelling block (eg 'there's a rat in separate')
- create a mental picture of the word
- write the word down from memory
- tackle only small groups of problem words at any one time; a mental alarm bell should go off each time one comes up, signalling the need for special attention
- check the word whenever it occurs; don't be afraid to check spelling by asking colleagues, or by being seen to use a dictionary
- learn spelling rules (eg 'i' before 'e' except after 'c').

Vocabulary

The English language is a wonderfully precise tool, and a good vocabulary is a great help in expressing ideas and concepts clearly and accurately. It is also an aid to brevity; one appropriate word can do the same job as several more general ones. However, English is also complicated, and has many traps and pitfalls. For example, the large number of words which sound the same but have completely different meanings.

Words frequently confused

It is easy to confuse words that sound the same but have different meanings (eg accessary/accessory; dependant/dependent). These words are used in normal design consultancy transactions, so it is worth giving them special attention.

Extending vocabulary

Make a committed effort to learn new words; this can be done in two principal ways:

- deliberately look up or question any word read or heard where the meaning is not understood. There is no shame in asking, and it is potentially far worse to pretend to understand, when difficulty may arise later because the meaning was not understood. This approach is also a valuable collective defence against jargon and the use of over-complicated language where it is not necessary.
- adopt a word-learning quota, and a specific time in the week for doing this. For example, a target of ten new words learned every Sunday evening would give a vocabulary extension of over 500 words per year.

Schoolroom grammar revisited

Before moving on to written style, business communications and the skill of expressing abstract ideas in words, it is necessary to establish the ground rules. The following notes are intended as both a quick refresher and as a reference guide.

Parts of speech

Noun a naming word: personal names (*Liz*); things (*book*); qualities (*mercy*).
Pronoun an alternative to the noun, eg *her; it*; used to avoid repeating the noun.
Adjective a descriptive word, attached to a noun or pronoun, eg a *difficult* task, a *striped* cat. Avoid being over generous with adjectives.
Verb a doing word, covering both action and states of being.
Adverb usually attached to a verb, to give more information; how/where/when/why is something happening, eg she worked *slowly*.

Conjunction a joining word, used to link words, phrases or clauses, eg *and, but.*
Preposition a word defining place in relation to a noun or pronoun, eg *to* Liz, *in* the book, *before* the flood.
Interjection an exclamatory word or phrase, often used in speech, but usually avoided in written work, eg *well, oh dear.*

Basic rule for sentences

Every sentence should have a subject, a verb, and an object, and the subject governs the verb. Long complicated sentences are probably too long; try splitting them into shorter sentences, or breaking them up with a semi-colon, as in this example.

Basic rules for punctuation

It is worth re-learning punctuation skills, partly because language changes and develops, and some rules change (often for the better), but also because ability in this area affects both the designer's own written output and his or her ability to help the client with copy preparation.

Full stops mark the end of sentences:

- each sentence should be a complete unit of sense which can stand on its own.
- the full stop is used to indicate an abbreviation but it is essential if the shortened form does not contain the last letter of the word, eg *Co., Etc., a.m.* take full stops but *Mr, Dr,* don't
- full stops are often omitted from acronyms, eg TUC, DBA, and this is often the preferable route for aesthetic reasons. It is important to be consistent in applying whatever decision has been adopted throughout the whole piece of work.
- items in a short list, introduced by dashes, do not need full stops.

Capital letters are frequently over used. They should be used:

- at the beginning of a sentence
- at the beginning of a passage of direct speech, quoted in inverted commas; eg *Mother said 'Come here at once' and I ran away*
- for proper nouns, eg *Design Business Association; Vicky; Mount Everest*
- for adjectives which have been made out of proper nouns, eg *Edwardian.* There is a tricky exception to this one, which is where the new term has passed into normal language and lost its link with the original source, eg *venetian blinds.*
- at the beginning of a line of poetry, except in some twentieth-century poetry
- for the pronoun 'I'
- when a noun is made into a person or expressed as a grand abstract idea, eg *Time ravages Youth*
- for references to deities and royalty, and any associated pronouns, eg *Buddha and His followers*
- in titles, for the first and all main words, eg *Professional Practice in Design Consultancy.*

Question marks are used in all direct questions, but not for reported questions.

Exclamation marks express surprise, sharp outburst or comment, and can also indicate humour or sarcasm. As a general rule, they are best avoided in business writing.

Commas should be used:

- to separate words, phrases or clauses in a list
- to break up sentences in a way which will make the meaning clearer, and the sense come across more easily. It is useful to think of a comma as the end of one thought, or a pause for breath.

Inverted commas may be either single or double; double are more commonly used for direct speech, but single tend to look better in text. The rules for using inverted commas are:

- for quotations (from speech, or from a book)
- for titles, such as the names of books or plays
- to indicate the use of foreign words, slang or unconventional terms (in print, italic can be used to do the same job)
- to separate a title within a passage of reported speech, eg *'Did you see "Cats" when you were in London?' he asked with some scorn.*

Semi-colons are an elegant and useful device, which can be used:

- to separate closely related clauses which could stand as sentences in their own right
- to divide up a list, eg *delegates should bring the following equipment: several pens; a loose-leaf file; a supply of good, white, unlined writing paper.*

Colons are used:

- to introduce a list (as in the line above)
- between a statement and an explanatory clause, eg *One thing is certain: we shall not surrender*
- to express a strong contrast, eg *John saves: Jane spends*
- to introduce a climax or concluding clause, eg *After pondering the choices before him, he came to a decision: he ran away*
- to make a pointed connection, eg *Humphrey became a board director in just three months: his father was the principal shareholder.*

Dashes and brackets are similar in function (and are mainly used to insert an afterthought) – or an explanatory comment – or a short list. They are particularly useful in adding variety in a long piece of writing, and help to achieve change of pace and clarity.

Hyphens can be used:

- to attach a prefix or suffix to a word, eg *multi-storey, President-elect,* and they should always be used in situations where confusion might result. Consider the differences in meaning between *re-sign* and *resign* (or even *re sign*).
- to form a compound word from two or more words, eg *day-to-day.* Compare *fifty-odd people* with *fifty odd people.*

This section is rounded off with some slightly longer notes on the *bête noire* of contemporary written English, the apostrophe.

Learning to love the apostrophe

The apostrophe is the most frequently misused element of punctuation; high streets are filled with signs to which unnecessary apostrophes have been added, and many greengrocers now sell 'tomato's' and 'potato's'.

Here is the simplest possible explanation of how to get it right:

- the apostrophe is used to indicate contraction of a word. It marks where letters have been left out, eg *did not = didn't; cannot = can't; they are = they're; I would = I'd; it is = it's.* Beware: *shan't* and *won't* are peculiarities.
- *it's* is not the possessive of *it*; but a contraction of *it is* (see above)
- the apostrophe is used for the plural form of certain letters and figures, but this is now dying out, eg *in the 60's; the three R's*
- the apostrophe is not used for possessive pronouns, with the exception of *one*; so *his, theirs, ours, its,* but: *one's* not *ones*

- the main use of the apostrophe is to show that something belongs to something or somebody, but unfortunately it is less simple than it sounds.

Using the apostrophe to denote possession

- decide who is the owner; establish whether singular or plural, and what is being owned, eg *the cat's whiskers* (one cat); *the cats' whiskers* (more than one cat)
- this gets a bit more complicated in the plural, eg *the sheep's feet* (one sheep); *the sheeps' feet* (more than one sheep – nasty); but *the baby's cot* (one baby); *the babies' cot* (more than one baby)
- with words ending in 's', the general rule is to add an apostrophe and a further 's' (singular) and an additional 'es' and then an apostrophe (plural), eg *St James's Street; The Jameses' house; Prince Charles's vegetable patch*
- there is one final variant, where the word ends in a sounded 'es'; here the apostrophe can simply be added after the word, eg *Mr Eliades' car.*

Written communication in business

In design consultancy, written communication is likely to fall into the following principal categories:

- informal messages
- note-taking
- business letters
- memoranda
- reports and proposals.

All the material in this section can also be applied back to the core business communications, with the aim of taking them beyond the efficient into the excellent.

Legal writing

It is important to mention that legal documents have been deliberately excluded from this section. This is for reasons which were expressed so clearly by Sir Ernest Gowers (see Gowers 1987) in his classic handbook that they are quoted here directly.

> *Acts of Parliament, statutory rules and other legal instruments have a special purpose, to which their language has to be specially adapted. The legal draftsman . . . has to ensure to the best of his ability that what he says will be found to mean precisely what he intended, even after it has been subjected to detailed and possibly hostile scrutiny by acute legal minds.*

However much one might wish to avoid legalese, it is extremely dangerous for lay people to attempt to produce legal documents. Designers can save time, and probably achieve a better working relationship with the company's solicitors, by thinking through the contents of a legal document carefully, making notes of objectives and requirements, but leaving it to the lawyers to translate it into a final written form.

Expert-to-expert communication

This section does not deal with 'expert-to-expert' communication. This is where two people share a field of detailed common knowledge, and can write to and for each other wholly effectively in terms which would come across as gobbledygook to the outsider. There are sufficient technical terms and elements of jargon in design practice to make this a real danger in communication between consultancy and client. Care must always be taken to establish the client's level of understanding at all stages of a project, and to write using only terminology with which they will be comfortable.

For example, relations between a large consultancy and a public sector client reached deadlock when it became impossible to reach agreement on fees and timescales for print work subsequent to a corporate identity project. Only after a third party had been called in as unofficial arbitrator did it emerge that the client did not understand the production process. Because of this, he did not understand the difference between 'design' and 'artwork', nor the time and costs associated with each process. Once this was explained, the problem disappeared and it was possible to reach agreement and move forward.

Informal messages

Because life in a design consultancy is fast moving it is important to establish basic ground rules for written message taking.

Every message taken should include:

- the date and time taken
- the name of the message taker
- the name and telephone number of the caller, with care being taken to check the accuracy of both the message itself
- a clear indication of what action is required.

When asking people to do things within a consultancy, it is often preferable to write a note rather than just ask verbally. This may seem absurdly formal in a small company, or between close colleagues, but a written message is less likely to be forgotten or misunderstood than a snatched conversation.

Note-taking

It is a valuable habit to always head a piece of paper with a date and a subject before writing anything else. When preparing notes of a meeting, include names of all people present, and if anyone arrives later on, record this at the relevant point in the account of the meeting. In making notes, the following are useful guidelines:

- keep notes as short as possible
- note only important points or facts
- note all decisions
- note all action to be taken, with names or initials of person(s) responsible
- include any dates or deadlines
- try to adopt a standard format for note-taking, eg A4, Filofax page, reporters pad; so that all notes can be filed easily
- lay out the notes carefully (bear in mind the 'going under a bus' principle; somebody else may need to use them unexpectedly)
- use main heading, subheadings, lists and numbered points
- use charts or diagrams wherever appropriate
- read through the notes as soon as possible after taking them, and if necessary refine or improve them
- consider the purpose of the notes; is it necessary to circulate them to other people, or to transfer material into a diary for future action?
- notes of a meeting (internal or external) are often written up as a contact report and circulated to all concerned as a record of what was discussed and agreed
- formal meetings are recorded through minutes which are not dealt with in this section, but there are several good text books on running meetings.

Another valuable habit is making what the law calls 'contemporaneous notes'. This means writing down what happened as soon as possible after it took place. The approach is especially useful after a difficult telephone conversation with a client. Contemporaneous notes are likely to be more accurate than a record made later, and even carry weight in legal disputes; this is one of the reasons why police officers carry notebooks. Notes are immensely valuable. It is neither wise nor possible to rely on memory for everything. Notes release space on the brain's hard disk for other activity.

Business letters

It can be taken for granted that design companies will have put considerable time and effort into the creation of their stationery; there will almost certainly be a grid layout for letter writing, and this should be followed at all times. It contributes greatly to the overall impression of a company from outside if all letters took the same, and, if possible, achieve a similar standard of clarity and helpfulness.

Key elements common to all business letters

It is essential to be clear who the letter is going to; check name, position and all spelling (with particular care over pitfalls like Francis (male) and Frances (female)). The recipient's full address should always be included in the letter layout; this creates a formal record of where the letter was sent, and is also useful for file purposes.

Salutation and endings

The beginning and end of a letter should always match. There is now much more flexibility as to how to open letters, but the recipient's first name alone should not be used unless he or she is an established friend or regular client.

The most common salutations and their respective endings are:

Dear Sir/Sirs/Sir or Madam	Yours faithfully
Dear Mr Da Vinci	Yours sincerely
Dear Leonardo Da Vinci	Yours sincerely

If the letter is more complicated in this regard, eg to a member of a royal family, it is worth consulting a book on contemporary etiquette.

Some people like to top and tail letters, that is to write in the salutation and ending by hand. This can add a personal touch to the letter, and allows flexibility. It also offsets the false personalization of letters which is now easily achieved with mail-merge technology (eg '*when, Mr Da Vinci, your gleaming new Batmobile arrives at Helicopter Buildings, N5*').

Your ref/our ref

If there has been a preceding letter with a reference, this must always be quoted. Whether or not reference numbers are used is a matter for individual companies to decide; they are useful, but can introduce an element of impersonality.

Date

Every letter must be dated.

Sender's name and job title

This should appear in typescript under the signature at the end of the letter. If the letter is *pp'd'*, the person signing should sign his or her full name, and put *pp* beside the typed name of the sender eg:

Jane Secretary
pp Leonardo Da Vinci
Managing Director

Do not *pp* letters if it can be avoided. It gives the recipient the impression that he or she is not of particular importance to the writer.

Enclosures

If the letter will be sent with other documents or items, this should be stated either in the main text of the letter, or at the foot, for example:

encs (3) – floppy disk
– photograph
– c.v.

An example of a standard business letter appears on page 62.

Useful techniques in constructing business letters

- state the subject very clearly at the start of the letter, or use a subject heading
- if the letter extends beyond one A4 page, check that it really needs to be this long: it probably doesn't, or perhaps a meeting is needed rather than a letter
- if a letter must go on to another page, mark 'continues' at the end of each sheet and number the subsequent pages
- order the contents of the letter clearly and logically, ending with any request for action, making sure that both subject and timescale is very clear to the recipient
- start a new paragraph for each topic or idea
- if the letter is chasing someone who hasn't replied, include a copy of the previous letter to avoid any further waste of time if the recipient claims never to have seen the original
- if replying to a communication, refer to it directly, eg *Thank you for your fax from Hong Kong, received 14 February 1992*
- when a letter refers to another person, eg *I have asked Mary to come to the meeting next Monday* it is usually helpful to send a copy to that person; the original, in this example, would be marked *c.c. Mary Gray* and all recipients would know the full circulation of the letter
- sometimes it is decided to send a copy to another person without notifying the addressee; this is known as a blind copy, and would usually be sent with a note saying *blind c.c. – for information.*

Style in business letter writing

Business letters should always reflect:

- the style, culture and standards of the company from which they come
- the nature of the relationship between the sender and the recipient.

Even where a relationship between consultancy and client is informal, correspondence should retain a degree of formality without becoming strained or artificial; this becomes easier with practice. Informal elements, such as *Hope your wisdom tooth extraction wasn't too bad* can be added in handwriting, to include a personal note in an otherwise official piece of communication.

Formal correspondence should always avoid waffle of all kinds. Resist the temptation to think on paper by making notes of the main points of the letter in advance. It is particularly important to do this before dictating letters.

Eliminate, as far as possible, all unnecessary words and phrases, and check for any possible ambiguity in meaning which could lead to a misunderstanding. Also check letters for overuse of particular words, and substitute alternatives where necessary.

When it is necessary to write a 'difficult' letter, preparation is particularly important. Marshall carefully the points which have to be made in the letter, and the objectives to be achieved. Never write and send a difficult letter in haste. By all means draft it while the issues are fresh, but leave the draft at least overnight, and then review it before sending. It may be helpful to take a second opinion on the draft from a colleague, and certainly to undertake role reversal, transferring the position from that of sender to that of recipient. Does the letter do its job? Is it adopting the right tone? Is it clear? Does it avoid self-indulgence, wild claims and gratuitous nastiness? Sometimes, writing a furious letter can be a great therapy (as long as it stops there) and the first, angry, letter is then replaced by a second, more considered version, which is the one that actually gets posted.

Memoranda

A memorandum (singular), memo for short, is principally a formal record of a meeting or conversation. In theory, a memo should not be used to request or initiate new action, but in practice, there is no reason against this as long as this route is clear and suited to the task in hand. (See page 63 for an example of a memo.) Normally, memoranda (plural) are sent out within a company, not externally, to:

- assist the sender by creating a formal record of his or her recollection of an agreed course of action
- make the information available to the other people directly involved
- make information available to other people indirectly involved or put people on notice of what is planned
- initiate action (and prevent words and good intentions from evaporating).

A memo is used when the issue in question does not require a longer document, eg a report, minutes, a discussion paper, or a formal recommendation.

Elements of a memo

- from whom, date
- to whom: this could be quite a long list, with the recipient's name underlined or highlighted; it is important that people receiving memos know who else has been sent copies
- 'copies for information': if people are indirectly involved with the subject of memos, (eg they were mentioned at the meeting or the decision affects their area of work) a copy of the memo should be sent for information; this is recorded as part of the overall text of the memo, so that everyone knows what has been done
- subject

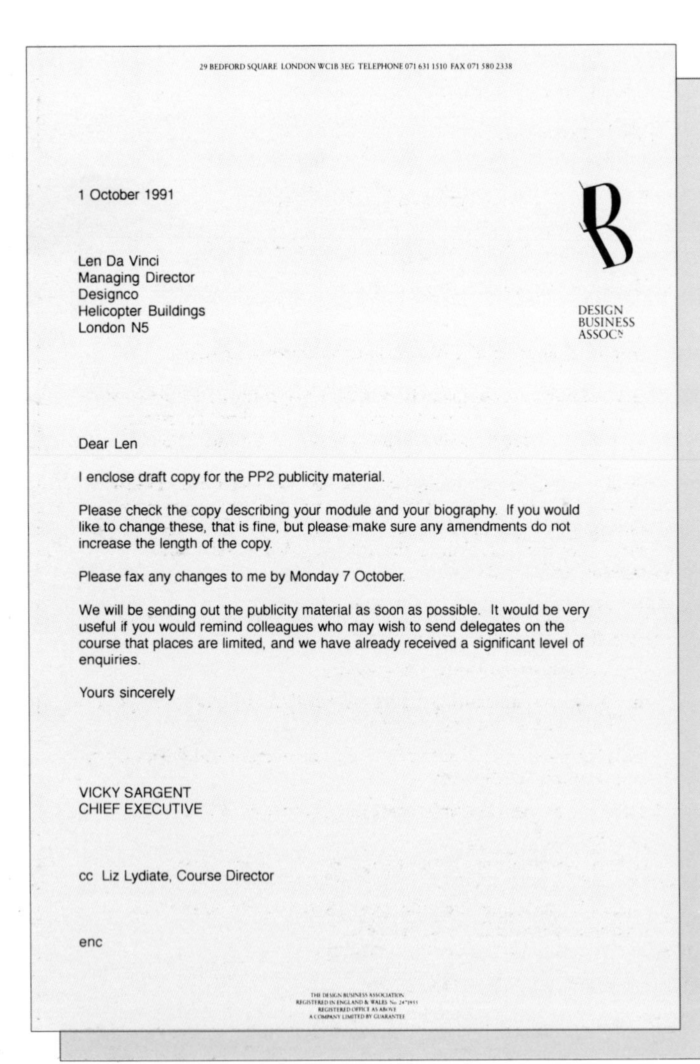

29 BEDFORD SQUARE LONDON WC1B 3EG TELEPHONE 071 631 1510 FAX 071 580 2338

DESIGN
BUSINESS
ASSOC

1 October 1991

Len Da Vinci
Managing Director
Designco
Helicopter Buildings
London N5

Dear Len

I enclose draft copy for the PP2 publicity material.

Please check the copy describing your module and your biography. If you would like to change these, that is fine, but please make sure any amendments do not increase the length of the copy.

Please fax any changes to me by Monday 7 October.

We will be sending out the publicity material as soon as possible. It would be very useful if you would remind colleagues who may wish to send delegates on the course that places are limited, and we have already received a significant level of enquiries.

Yours sincerely

VICKY SARGENT
CHIEF EXECUTIVE

cc Liz Lydiate, Course Director

enc

THE DESIGN BUSINESS ASSOCIATION
REGISTERED IN ENGLAND & WALES No 2471955
REGISTERED OFFICE AS ABOVE
A COMPANY LIMITED BY GUARANTEE

- information conveyed/activity recorded/action requested; this should be concise, and may be set out in note form, with numbered points
- memos may be marked 'confidential' if appropriate; in this case they must be sent out in envelopes. Non-confidential memos are usually distributed without envelopes.

When not to use a memo

Care must be taken not to use memos inappropriately, particularly in circumstances where there are sensitive issues which may require different treatment. For example, the memo shown here, referring to research about CAD equipment, could give rise to anxiety amongst staff about restructuring or even redundancies. Because a non-confidential memo is a piece of accessible internal communication, always think through the consequences of information leaking to a wider audience.

Memos also should not be used to introduce issues or communicate information which could be better handled through a meeting. A memo could record the outcome of such a meeting.

Reports and presentations

This next section offers help with the preparation of long, self-contained documents in project handling, and the following material concentrates on general principles in compiling and writing reports and proposals.

Documents of this kind should be complete in their own right, and tell the whole story. Any tables, references or other material which the reader will need should accompany the document as appendices. The document cover should explain who the report is from; who to; who wrote it; when and the subject.

Notes on the various stages in preparing and writing a report are set out below.

Preparation

- collect together all material which might be relevant; rough notes, xeroxes, reference books, correspondence, and newspaper cuttings
- go through the material and look for a structure for the report; this can follow the army model of 'say what you're going to say; say it; sum up what you've said'
- work out where each piece of information might fit within the structure, beginning to decide what should be included and what should be discarded

Confidential

MEMORANDUM

From: Chief Executive (or Len Da Vinci)

To: Production Director (or Mike Buonarroti)

Date: 3.10.91

Proposed adoption of CAD capacity for artwork division

As you know, the company is considering opening a new artwork division in the near future. Our operations have been expanding so rapidly that a development of this type now seems inevitable.

In addressing this possibility, we now need to review our options in terms of CAD systems, capable of handling all in-house typesetting and computer generated artwork. I would be grateful to receive your written recommendation on the following, before the Board meeting on 1.11.91:

i) Which type of system do you recommend for our purposes, and why?

ii) Should we buy or hire equipment?

iii) Can we retrain existing staff, or will we have to create new posts? What are the cost implications?

iv) What provision should be make for maintenance, and back-up in case of systems failure?

v) What are the cost implications of iv), in terms of loss of production?

vi) What is the expected life of your recommended system?

I look forward to hearing from you; do ring me if you think a preliminary meeting would be useful.

- review the structure in terms of the original brief; is the work so far to the point and dealing with the right issues in a clear and effective order? It is easier to change the structure at this stage than later on, when more work has been done and there may be a greater emotional commitment to maintaining the status quo.
- if appropriate, discuss or check the structure with any relevant colleagues.

First draft

- start writing; this is putting flesh on the bones of the structure. Develop each theme, starting a new paragraph for each point, and making sure to link each paragraph to the preceding and succeeding ones.
- use straightforward language and avoid over-long sentences; vary the length and pattern of the sentences
- choose words for their accuracy of meaning, and appropriateness in context
- avoid using several words where one would do
- don't write anything which is unnecessary; avoid repetition, and don't ramble
- avoid being pompous or over-formal, but also don't use slang or colloquialisms
- if it is necessary to include an opinion, this should always be backed up by reasons
- consider the use of headings, sub-headings, paragraph numbers, bold/italic/capitals etc.; this is an editing job, to present the material in the clearest manner possible
- move sections of the draft text into a different place, if the initial structure can be improved upon; word processors were made for this, but if necessary scissors, Pritt, Sellotape and a Xerox machine are also helpful in drafting reports
- remember that it is possible to write around a hole if a piece of information is needed or awaited. Also, if one section becomes difficult, it can easily be left and dealt with later. Think of the units of the draft as blocks which can be moved around independently.
- prepare the draft as a typed or word-processed document, set out in double spacing with wide margins.

Correcting the draft

- leave time between preparing the draft and making the corrections, so as to approach the piece fresh, and to consider how it comes across to a new reader
- refer back to the original brief/objective
- work systematically through the draft, looking for:
  - bits which aren't clear
  - bits which could be expressed more succinctly
  - places where extra information, facts or references would help
  - places where the structure might be improved
- check spelling/accuracy/sense/overall clarity of the material
- review headings, paragraph numbering and pagination
- mark corrections clearly in a second colour.

The final version

- print out the corrected draft, and check it over again
- circulate the draft to relevant colleagues or anyone else involved, inviting their comments by a set date; it is a good idea to cover this stage with a memo, so that it is clear who has been consulted
- incorporate any changes, and produce the final text; at this stage, destroy early drafts but retain any marked-up copies of the final draft which have come in from colleagues, in case any questions arise

- consider whether the document would benefit from adding a contents list, a summary or even an index of key points
- look at whether it is possible to make life easier or more interesting for the reader by adding supplementary material such as illustrations or charts
- give attention to how the document will be presented
- remember to circulate finalized copies to everyone who has been involved in the preparation, and to all key players who need to be kept informed on the project.

Summary

It is of course impossible to offer more than a starting point and some general signposts in a chapter of this nature. However, if these guidelines do succeed in offering a way forward, and some encouragement, they will have met their objective. Language, and in particular written language, is about progress and development, and it is also an immensely versatile tool. Design practitioners can make it their own, and it will work happily alongside pen, pencil and computer.

Further reading

Crystal, David (1985) *Who Cares About English Usage?* London: Penguin.

Gowers, Ernest (1987) *The Complete Plain Words.* London: Penguin.

Hayakawa, Samuel (ed.) (1991) *Cassell's Modern Guide to Synonyms and Related Words.* London: Cassell.

Littlejohn, Andrew (1988) *Company to Company.* Cambridge: Cambridge University Press.

Reader's Digest (1986) *The Right Word at the Right Time.* London: Reader's Digest.

Roget, Peter Mark (1990) *Thesaurus of English Words and Phrases.* London: Penguin.

'It is not the
failure of others
to appreciate
your abilities
that should
trouble you, but
rather your failure to
appreciate the abilities of other
people.'

Confucius

Mick Armson/Sharp Practice

7 MANAGING YOURSELF AND OTHERS

Shan Preddy

This chapter provides an insight into some of the basic skills involved in managing the people working in design companies, and deals with:

- management theory
- assessing strengths
- motivating others
- setting goals and standards
- reward and punishment, praise and criticism
- managing groups and team leadership
- delegation skills.

Much of the material links into chapter 14 which discusses the issues involved in running a design consultancy.

What is management?

Why is it so important to learn the necessary skills involved in successfully managing yourself and others? In design consultancies, it is often the case that talented and able employees, who have proved themselves to be excellent at their work – whether in design, marketing, client service, production or administration – are either promoted to the position of manager, or tempted to leave in order to set up their own business. Suddenly, they are expected, without any formal training, to manage one or more other people, who have up until that time been peer group colleagues. In addition, they are expected to carry on producing the same excellent work which gave them the recognition in the first place. Naturally, some people find the transition difficult. Moreover, despite years of experience, some owners, partners and directors of design consultancies never learn how to manage and motivate their staff properly; as a result their companies are not as successful as they could be.

Good managers are not necessarily good leaders; some people are talented at organizing others without having the vision, respect and commitment required of leadership. However, a good leader will not succeed without also being a good manager. Technical knowledge, in one or more specialist fields (such as design or marketing), is a necessary and desirable attribute of a successful manager: in terms of gaining respect from others; in understanding the tasks which others are asked to do; and in setting standards and examples. However, this knowledge alone is not enough. The successful leader must also have considerable skills in people management.

The purpose of a manager is to achieve results through other people, while following established company policies and directives. Results may be achieved through additional work of the same type; through work which uses talents and skills which the manager does not possess; or through the delegation of administrative or other tasks which otherwise prevent the manager from tackling his or her own responsibilities. The purpose of an employee is to achieve improved results for the company, better results than the employer could achieve alone. If those improved results cannot be achieved by employing one or more additional individuals, the employer would benefit from working alone.

The theory is simple enough. Problems start for most managers when trying to put the theory into practice. Failure results when:

- the other people through whom the results are to be achieved do not also want to achieve the same results, or
- they do not know which results they are supposed to be achieving, or
- they do not know how to achieve those results, or
- the manager does not know how to achieve those results.

Assessing strengths

It is important that any manager of people is clear about two fundamental aspects of any project.

1 The objectives of the task, or project, or company

What is this individual or group of individuals trying to achieve? The objective might be very straightforward, such as: the completion of Stage 1 design concepts to a degree of finish which will allow them to be discussed internally at a meeting at 3pm on Thursday. Alternatively, the objective might be complex, and far-reaching in its implications for the whole company, such as: by 1999 we will be a 'top fifty' design consultancy specializing in the leisure sector and operating on a pan-European basis.

Once the objective has been set, a strategy for achieving that objective can be developed. Without a clear objective, or statement of expected results, no manager can hope to succeed. William of Orange, a royal prince, sailor and tradesman later to be crowned William III, said in the seventeenth century: 'No wind is favourable to a man who does not know where he is going.' Design professionals, like everyone else, tend to blame external factors for the failure of a project, when the real failure often lies with the lack of clearly defined objectives. The later section in this chapter on goals and standards gives further information in this area.

2 Assessment of relative strengths and weaknesses

The relative strengths and weaknesses of both the manager, and of the people reporting to that manager, need to be

A SWOT analysis for a junior designer

Strengths
- good, creative ideas
- enthusiastic
- willing to learn
- fluent French
- hardworking
- gets on with rest of company

Weaknesses
- inexperienced
- poor attention to detail
- no Mac knowledge
- low technical ability
- poor time-manager
- learner driver
- poor presenter

Opportunities
- Mac training
- time management training
- presentation skills training
- Paris office
- get driving licence

Threats
- may get bored
- may leave once trained
- may be unable to cope with computer training

assessed. People working in design, at whatever level of seniority, can benefit from defining their own strengths and weaknesses, relative to other design professionals. This definition can then form part of a career path. Although the design industry does not lend itself to a straightforward career pattern, as might be experienced in other fields, it is nevertheless helpful to be able to use talents to their best effect. Obviously, this can only be done if those talents are recognized.

An assessment can easily be made, using the SWOT analysis technique, in which the relative strengths, weaknesses, opportunities and threats relating to the manager or the individual are placed under the appropriate headings. Once identified, it becomes clear that all four apparently contradicting areas are, in fact, closely related. As a demonstration, an example of a SWOT analysis for a junior designer is shown opposite on page 68.

In this analysis the individual clearly has talent and potential. Some design and personal skills training seem appropriate, but an incentive to stay with the company once that training has been completed might be required. Perhaps the promise of a period spent in the Paris office might prove attractive. Poor attention to detail and low technical ability could mean that computer training could be slow, and that extra management supervision will be necessary on any projects undertaken. However, the good creative ideas will be beneficial. Perhaps joining forces with another designer who is good on the computer and at detail, but lacks ideas, would be an insurance.

An analysis like this can be done very swiftly, and can provide a firm guide both to one's own abilities, and to the development potential of a given individual. A team of people can then be established to work on a particular task. (Information on team leadership appears on page 73 to 74.)

Motivating others

Creating a team based on individual strengths is only the beginning. Once the team exists, it needs to be motivated.

In order to achieve results through others, it is often necessary to encourage someone to do something which they might not want to do. It may be that a client needs convincing that a particular design route is the best one to follow; it may be that a junior or a colleague must be asked to work late; a supplier might need to be persuaded to drop a price; the boss needs to be approached for a pay rise.

Many people are motivated by things other than money, in life as well as at work. A good manager of people needs to be able to identify what will motivate a given individual under a given set of circumstances. There are many different ways to get someone to do something, or in other words, motivate them. Not all of them will work in the long run.

Some are negative influences: physical force; bribery; blackmail; threats; the creation of guilt; flattery; or the presentation of a *fait accompli.*

Some are positive influences: praise; reward; promise; transferring ownership of an idea; obtaining moral commitment; getting approval in principle; involving people in the decision-making process; seeing it from the other person's point of view and telling them so; creating a challenge or a competition; or setting them a reputation to be lived up to.

However, the best way of all is to make the proposition attractive. Make the other person want to do it. How can this be achieved?

The commitment gap		
Response	**Problem**	**Solution**
'Can't'	Ability	Training
'Won't'	Attitude	Management

- Concentrate on the other person and their individual needs, wants, hopes, desires, fears, feelings, beliefs. How do they feel about this particular situation? Most poor motivators concentrate instead on themselves and what they want to get done.
- Point out the benefit to that individual in agreeing to the task or the suggestion. What do they stand to gain as a result of carrying out this task or agreeing to this suggestion? What is in it for them?
- Make it easy. People object to a course of action when a series of apparent difficulties emerge. If difficulties are removed, the objections also vanish.

If all of this sounds like sales talk, it is. Whenever someone tries to persuade someone else to think, feel, believe or do something, it is the 'sale' of certain a point of view. (See also chapter 5.)

Sometimes, no matter how much effort has been spent on working on their motivation, people still do not always accept a point of view, or agree to do something. It may be that they are genuinely unable to do it, or it may be that they are reluctant to do it for some other reason, such as lack of commitment to the task, the manager or the company. The verbal response will almost always be variations on 'I can't', rather than 'I won't', but the real reason may be closer to a refusal based on attitude rather than on genuine inability. Careful questioning of the individual should lead to the answer.

A genuine 'can't' response is often related to time pressure. In this case the solution is either a management decision on that person's workload, or specific time management training to combat the problem in the long term.

Goals and standards

One of the fundamental principles of managing others is setting clear goals and standards for all individuals. For the manager at the top of the organization, goals should be set by the rest of the senior management. Without clearly defined targets, people do not know what is expected of them. Equally, a manager cannot accurately assess an individual's performance if criteria for assessment have not been established.

A goal is a statement of results which are to be achieved. An example of a goal might be: 'To be responsible for work which will win or be shortlisted for a Design Business Association Effectiveness Award for print work in two years' time.'

A standard is a measurement of ongoing performance criteria which must be met continuously. For example: 'To be at desks and working by 9.00am every weekday morning, unless out on company business.' Of course, once a goal has been achieved, it often becomes a standard for that particular individual.

Both goals and standards should be:

- ***set jointly by the manager and the individual:*** if goals are imposed on an individual by a manager, there is likely to be little commitment to achieving the desired results
- ***specific:*** statements like 'telephones are to be answered quickly and messages taken when necessary' sound efficient, but are not specific enough to be useful. How quickly is quickly? When is message-taking likely to be necessary? A more specific statement of a standard would be: 'Telephones are to be answered after no more

than three rings. The telephone is to be answered in the manner outlined in the company handbook.' Specificity is easier for some jobs than for others. A marketing or sales person can have financial or volume targets as goals. For a designer, goals are more unstructured, but it is important to be as specific as possible at all times.
- ***achievable:*** goals and standards should be realistic. If they are not achievable, the individual will lose motivation. However, goals should also be challenging: intelligent employees respond well to being stretched within their capabilities.
- ***measurable:*** if the goal or standard cannot be measured, how can the individual or the manager know whether it has been achieved? It is also important to specify how the measurement is to take place, how often, and by whom.
- ***frequently assessed:*** there is little point in assessing the achievement of goals once a year, at salary review time. The manager and individual should talk frequently about progress, with the manager directing and helping the individual towards success.

Reward and punishment, praise and criticism

The ability to use reward and punishment, praise and criticism, is another fundamental management skill. Reward, thanks and praise create a successful working relationship. Punishment and criticism are necessary, but, if used incorrectly, can destroy that relationship.

Constant nagging, negative criticism, or a general lack of appreciation for effort will cause people to feel negatively towards their manager, and will quickly reduce both specific and general motivation in that individual.

The cry of 'lack of appreciation' is something that runs through most organizations, almost every design company, and through many personal relationships. Many talented and otherwise valued staff move to a new job because they feel that their skills are not being appreciated.

Praise and reward

The distribution of praise or the giving of a reward of any kind is meaningless if not handled properly. Praise and/or rewards must be:

- ***specific to the work in question:*** general praise along the lines of 'good work, carry on… ' is pleasant, but pointless. After a while, it will be taken by those whom it was intended to motivate as the empty phrase that it is. Praise, or a reward, should be attributed to a specific piece of work, or action, or result.
- ***personal to the individual concerned:*** the individual who is receiving the praise or the reward should be named, and spoken to as an individual. A manager who walks into the studio and calls out 'well done everyone, the client liked the work' is, by talking to everyone, not talking to anybody in particular. Some members of that group will have worked harder than others, or had better ideas. Recognition for individual effort is vital to motivation.
- ***given in public:*** private praise or reward is motivating; when given or announced in public it is even more so. In addition, others in the company will want an equal amount of praise or reward, if they see that it is meaningful.

Criticism and punishment

In any relationship, work or otherwise, there is occasionally the need to tell people that something they are doing or have done is not acceptable. If behaviour is not stopped, it is reinforced. As the old phrase says: 'Silence will be deemed to be tacit acceptance.' A lack of reprimand will be deemed to be tacit approval. However, a badly given reprimand is damaging. Any reprimand or criticism of another individual's work or behaviour should:

- ***be specific:*** general criticism will be taken as general nagging, or general irritability; criticism of a specific piece of behaviour will be absorbed
- ***separate the behaviour from the person:*** when reprimanding someone, it is important to communicate that it is the specific behaviour which is unacceptable, not the whole person. If phrases such as 'you are lazy', or 'you are a terrible designer' are used, the immediate reaction from the individual criticized will be one of defence. They will stop listening to the specific reprimand. Instead, the behaviour should be discussed in such a way that both parties can view it objectively. Phrases such as 'that report was very late' or 'the last project you completed was well below your usual creative standards' are easier to accept. They are still criticisms, but the recipient will be able to comment on the behaviour without feeling as though an attack is being made on a personal basis.
- ***be private:*** reprimands should be given quietly, in private. This is not suggested as a method of sparing the individual's feelings (although in some circumstances this may be a valid consideration) nor as a failsafe device in the case of the manager being wrong about the facts involved. It is suggested for one simple reason. If a manager criticizes another person in the presence of others, the remainder of the group will almost always, regardless of the rights and wrongs of the situation, take sides with the person being criticized, and not with the manager.
- ***be immediate:*** giving a reprimand long after the event in question will only encourage scepticism about the seriousness of the error or unacceptable action. If the issue is important, it must be tackled immediately. If it is not important, it should be forgotten. Reprimands should not be given in anger. A little fake anger can help to make the point, but real anger will not succeed. It can cloud the ability to think clearly and rationally, and can also lead to unfair accusations. Instead of giving a reprimand in the heat of anger, wait until a later, suitable opportunity.

The wording of any reprimand or criticism is important. If the word 'and' is used instead of 'but', the recipient will be more motivated to do any additional work required. For example, 'Your report on the question of workload was good, but completely ignored the issue of costs. Could you revise it?', might be better expressed as 'Your report on the question of workload was good, and would have been improved by the addition of a section on costs. Could you revise it?'.

If people are very sensitive to criticism, attention can be drawn to similar mistakes made by the manager concerned in the past; people like to feel that they are working for a human being, rather than an infallible robot. However, excessive mention of mistakes made by management will destroy trust in its strength. A mature company is a learning organization. It allows people to learn by their mistakes. Once.

Managing groups

Managing a group of people is different from managing one person; the issues become more complex, and the hopes, fears, beliefs, opinions, feelings of several individuals are more complex. Individuals influence the behaviour of groups. However, groups resemble individuals in that they are all unique. Many managers have made the mistake of managing one group in an identical manner to another, in the belief that what worked once will work again. Sometimes it does, but it cannot be relied on.

Groups, or companies have a collective personality, whether that personality is positive and happy, negative and demotivated, hardworking and ambitious, or lazy and complacent. This personality (good or bad) can be created, developed, and formed by the person or people at the top of the group or company, in other words by the leader. This is particularly true when the group is in a formative stage.

Leadership is always best done from the front. It is a question of 'Do as I say and do' rather than simply 'Do as I say'. Setting examples (corporate, role or gender models) is a very powerful management tool.

Group and individual needs

There are three sets of needs facing any group of people, however large or small that group may be, and any manager has to learn to balance all of these demands. First, the whole group has needs, which may be very different from the needs of other, apparently similar, groups. These group needs are determined by the sum of the needs of the individuals making up that group.

The second set, individual needs, are very entrenched. People at all levels of seniority, whether a college entrant or an experienced owner or manager, need to feel that they are rewarded financially, that they work for a company with a good reputation, and (as far as is possible to foresee) a healthy future. Employees need to feel that they have a future career plan. Everyone needs to feel that they own their work, that they can see the results of their efforts, and be recognized as a contributor to those results. They need to feel that they are involved and consulted, and are kept informed of what is happening in the company. People need to be challenged, to think for themselves, to have the opportunity to develop skills, and to learn.

Third, there is the need to achieve a common objective, whether the task in question is the future success of the company, the completion of a project, or a decision on where to go for lunch.

Any one of these needs affects the other two.

From a negative viewpoint:

- if the task fails the confidence felt by both the group and the individuals in that group suffers
- disunity or lack of harmony in the group affects the task and the individuals
- a problem with one of the individuals affects the group, and the task.

From a positive viewpoint:

- achievement of the task boosts group and individual morale and performance
- group success, or a good team spirit, affects individuals, and therefore the task
- individual contributions or acclaim help the group, and therefore the task.

The role of the manager of that group is, therefore, to try to achieve the task or established company objectives, to build and develop the individuals, and to maintain the group or team spirit and effort.

Team leadership

A loosely knit group of people becomes a team when they have an objective or common purpose which is clearly understood by all members, when each member has an assigned role to suit their own talents (see page 69), and when everyone is clear about those roles so that duplication, omission and confusion are avoided. When all team members integrate their talents, the shared objective can be met. There should be no demarcation by status; everyone is involved. Communication should be open, with the opinions and views of team members taken into account; they are valuable. Feedback from all parties should be rapid, and thanks and specific praise will motivate the team for the future.

The role of the team leader

Ideally, the team leader, whether a project manager or a managing director, should help to select the members of the team, be they senior or junior to the leader. On a design project, team members will come from the design consultancy, from the client company, and from relevant suppliers or contractors. The team leader should co-ordinate the team effort, ideally overseeing the project from start to finish, even if other team members change. The leader should make sure that all of the team members know and understand the overall objectives and strategy, and understand their own jobs, so that duplication of effort and omissions are avoided. The leader should work as a member of the team, creating commitment from members by understanding their needs and wants, by building a motivating environment, and by establishing the lines of communication between team members, and between the team and the outside world.

The leader should resolve any conflict within the group; provide a means of problem solving; be able to measure success and failure, and communicate the results back to the team; trust the team members and stand by them and support them when necessary; and give praise and reprimands where appropriate. It helps to be able to develop a reward system (not necessarily financial) for members.

Lastly, good team leaders, or group managers, know when to do the work themselves and when to delegate it to another member of the team, or to someone outside of the immediate team.

Delegation

Delegation is not about having someone else to dump work onto. It is about achieving results through other people, that basic management principle.

Poor delegation practice within an organization means:

- deadlines are missed, or narrowly met
- some employees are busier than others
- employees are unsure of their authority, and no-one knows who is in charge
- managers are too busy to talk to people when they need advice, and never have the time to walk about the company
- decisions and communications are slow, incomplete, and often too late
- managers intervene in projects and overrule decisions, so demotivating juniors
- talented employees get bored, and leave
- managers work late, take work home and wonder what it is all about.

Good delegation practice within an organization means:

- more work can be accomplished
- deadlines are met
- control of work becomes easier
- employees become more involved in, and therefore committed to, the work
- employees learn, grow, develop and become more useful to the company
- managers have more time for the tasks which only they can do: planning, organizing, controlling, motivating
- managers go home at a sensible hour.

Good delegation practice involves:

- having a sense of the value of time and capabilities
- knowing that it is not an abdication of duty
- deciding who is the best person for the task and not delegating those exceptional tasks which, in fact, only the manager can do
- giving a clear and comprehensive brief, setting specific, achievable and measurable goals and standards
- giving people the power to make decisions, and standing by those decisions; authority as well as responsibility should be delegated
- determining the level of support offered: training, resources, budget
- determining the nature and the frequency of the feedback required from others and provided to them
- making sure everyone understands who is in charge of the project
- leaving well alone, but being there if needed
- giving feedback, and credit where it is due.

Poor managers often complain that they are running out of time; in reality their subordinates are running out of work. Delegation is a great motivator; it stands for trust.

Counselling

Feedback is an important factor in successful communication between people who are working together. One method of formalizing feedback is in establishing an effective assessment, or appraisal, system (see chapter 18). However, from time to time, there will be a need for a semi-formal session between a manager and an employee, which is intended to resolve one or more issues which are not directly related to a specific project. These sessions are counselling sessions, and, as with appraisals, a counselling session can, when handled well, be a useful interaction with positive results. If handled badly, however, such a session can irrevocably damage an otherwise satisfactory working relationship.

Most requests for a counselling session will be phrased: 'Could I have a chat with you sometime about...?' and may be instigated by either the manager or the employee. Problems which result in the need or request for a counselling session can be large or small: colleague problems; career matters; stress, whether work related or not; redundancy, dismissal or demotion; work-related problems (for example a reluctance to work abroad); and, of course, personal matters.

Successful counselling requires certain skills, which can be learned. Above all, counselling provides a form of detached but knowledgeable listening, and will give a different perspective on the matter under discussion, so that the person with the problem gains a better understanding of it. The skills required are: active listening; questioning to clarify the issue (see chapter 9); encouraging the other person to generate options and encouraging the other person to analyse those options; decision-making (theirs, not the counsellor's) and a clearly given summary of next action to be taken by the individual.

The real art of counselling is to be a facilitator to a solution, not to be a source of all knowledge with an answer to everything.

When a manager or a colleague suspects someone has a problem which is affecting their work, it is dangerous to make assumptions. Instead, enquiries should be made, directly, but diplomatically. An example of an approach might be: 'You don't seem to be very happy at work at the moment. Is anything bothering you? Can I help in any way?' The problem that the individual discusses initially may not be the underlying problem. It is important, therefore, to create a relaxed and unhurried environment for the discussion.

Decisions should not be arrived at hastily; judgements on the validity of either party's case should not be made. Verbal and non-verbal signs of encouragement can help, as well as a demonstration that the counsellor is actively listening. Notes should not be taken: they will inhibit openness.

If the problem is personal, listen sympathetically, and suggest, if the problem is serious enough, that professional advice is sought. Examples here might include: medical advice (body or mind); marriage or relationship guidance advice; alcohol or other drug abuse advice; and financial advice. Offer help in finding advice; the employee has shown trust in disclosing his or her problems, and the counsellor should be prepared to offer some support. However, the counsellor should not take decisions for the other party; it could be damaging.

Stress

One common subject for counselling is stress. The design business is frequently cited, by design professionals, as a stressful business. Compared with some occupations, such as emergency medical teams, it is, of course, relatively stress-free. However, it is a high-pressure business, with short deadlines and the need to juggle several different assignments at the same time. Specific influences, such as recession or lack of business, create different sorts of pressures.

Pressure is often what makes people get things done, and everyone can work under surprisingly large amounts of pressure for short periods of time. It is when that pressure turns to stress that problems occur. Mistakes happen, work takes much longer than it should, people start to feel that they cannot cope. Like a pair of scales, it just takes one thing too many and the balance is lost. If anyone is showing signs of real stress, such as sleeplessness, memory loss, unusual irritability or other personality change, unwarranted tearfulness, depression, or if physical symptoms occur (such as severe headaches or unattributed back pain), medical advice should be sought.

In all cases, whether the subject matter is work-related or personal, the counsellor should listen. When the person seeking help acknowledges the problem, they should be asked what they think can be done about it. Only then should a counsellor offer advice, where appropriate. A course of action can then be planned together, and a follow-up meeting arranged. A good counsellor will always lead the other person to make the decisions.

Conclusion

The theories of managing other people briefly outlined in this chapter are, in themselves, simple and straightforward. As mentioned at the beginning, it is the practice that can be difficult. As one very experienced manager said: 'Right, that's me sorted out. I understand how to manage people much better now. The only trouble is, what do I do with all of the rest of my company who haven't studied the theory?' The answer is, of course, that the rest of the company needs to be convinced, by the instructions, the behaviour and the example of the manager.

People are the prime assets in any design consultancy; they are also the most expensive items on the balance sheet. Moreover, they are assets which are allowed to walk out of the door every night. As with all assets, it is worth making the effort to manage them properly, and to do everything possible to ensure that they will choose to return the following morning, to a happy and productive working environment.

Further reading

Adair, John (1988) *Effective Leadership.* London: Pan.
Blanchard, Kenneth and Johnson, Spencer (1987) *The One-Minute Manager.* London: Fontana.
Carnegie, Dale (1986) *How To Win Friends and Influence People.* London: Heinemann.
Dell, Twyla (1989) *How To Motivate People.* London: Kogan Page.
Drucker, Peter (1988) *The Effective Executive.* London: Heinemann.
Fletcher, Winston (1990) *Creative People: How to Manage Them and Maximise Their Creativity.* London: Business Books.
Maddux, Robert B (1988) *Team Building.* London: Kogan Page.
Maddux, Robert B (1990) *Delegating for Results.* London: Kogan Page.
Stemp, Peter (1988) *Are You Managing?* London: Industrial Society Press.

Jamel Akib/Sharp Practice

'Perfection is the child of time.'

Bishop Joseph Hall

8 TIME MANAGEMENT
Liz Lydiate

Time management in any business falls into two areas; the behaviour of the individual, and the performance of the group. Much of the material in this chapter relates to the principles and techniques discussed in chapters 7, 9, 13, 14 and 18.

The organized designer

Most people attempt to divide their lives into work and non-work activity, but those in creative fields, including designers, tend to have a more realistic understanding that all areas of life are interconnected. The designer or design administrator has only one brain, and draws on its resources for all stimuli, facts and decisions. Consequently, experience achieved during non-work time will have a direct influence on work time performance, and vice versa.

Because the skill of the designer is the provision of creativity, it is essential to develop work patterns and principles which are suited to delivering work for clients, as effectively and efficiently as possible, but also give attention to the need to replenish, enhance and deliberately maintain the creative edge. A lot of tired, uninspired design is the result of misguided attempts by designers themselves, or management, to give out more than they put in, on an ongoing basis. Time management for design must therefore be based on the recognition and establishment of self-maintenance needs, and make adequate provision for:

- maintaining personal time
- input for greater creative output
- self-development.

These elements represent the foundation, onto which various time-management skills and techniques can be built. Despite any other external pressures, the designer has to recognize the central importance of caring for health/self/relationships. If these things are not right, there is a direct effect on the capacity for and quality of work.

Because of the interconnected nature of life in design, it is often possible to combine the meeting of a number of needs in one activity; for example, meeting a friend to visit an exhibition, joining the company softball team, attending an evening class to learn a language.

In examining the individual in relation to time handling, it is helpful to consider three basic concepts which fundamentally affect performance and decision making.

Guilt and overcommitment

People working in design are particularly prone to these problems, partly because most of the time the work is of itself attractive and pleasurable.

Recognizing that failure to deliver, on time, or to brief, is one of the greatest single sources of client dissatisfaction, is an important step towards overcoming guilt about leaving the studio at a reasonable time, or saying 'no' to a particular deadline.

Reading must be considered as an essential part of work, and time allocated accordingly. Perhaps it is possible to travel by public transport and read during this time, or to group books/articles/papers for attention on train or plane journeys. If travelling by car, consider using the time to listen to material on tape. A book carried at all times is a great protection against time wasted by delays, queuing and people arriving late for appointments.

Self-development only happens as a result of a deliberate decision to do something new; consider where and when time can be captured for tackling a specific project, and pursue this without guilt – it is an important investment for the future. Achieving a confident stance on the issue of input for greater creative output is easier once this is accepted as essential rather than desirable. This means overcoming short-term vision, which says it's more important to do something dreary and routine, rather than something stimulating and unusual. Guilt of this kind has to be overcome: 'creative batteries' can only run for so long without recharging. The person who works round the clock as a habit rather than an exception will quickly drain his or her resources.

Concentrating and single-handling

Design studios are busy and demanding places. Once proper decisions have been made about how time is to be allocated, it is vital to be able to make the commitment to addressing a single piece of work at a time, without worrying about other things which are consequently not being done. The human brain is not designed to concentrate on more than one thing at once, and most mistakes result from interruptions or attempting to do things too quickly.

Having prioritized tasks and allocated time accordingly, stick with the activity until it is finished, resisting all interruptions (perhaps using the techniques described elsewhere in this chapter). Never feel less than 100 per cent committed to the task in hand; relax and enjoy the work. The other things cannot be handled at the same time and they must not be allowed to poach attention allocated to something else.

Work which is identified as a top priority must be given this kind of prime time if the person doing it is to have a chance to perform well. In order to achieve this, calculate carefully the time needed, and then add 30 per cent. If this extra time is not needed, it can be recycled towards getting ahead on something else, which is another key technique.

Chapter 9 gives guidelines on handling requests which cannot reasonably be met; it is useful to remember that any individual cannot be expected to know or keep detailed track of another person's total commitments, and will often assume that if a request is accepted it is because the colleague is not overloaded and the work can be handled without difficulty. Similarly, there is no shame in asking for help, as long as all possible efforts have been made to address the problem and tease out possibilities before making the approach. Although everyone is busy, the 'stitch in time saves nine' principle applies, and colleagues will be much happier to have the opportunity to help at an early stage rather than take part in collective fire-fighting later.

Delegating

Designers are often bad at delegating, because it requires the acceptance of a principle which goes against the design ethic of individualism and perfectionism. This principle is that successful delegation involves accepting that the delegatee will carry out the task in his or her own way, and that although this may be quite satisfactory, it will not be identical to the way in which the delegator would have done it. The pursuit of identical performance is pointless and leads to great unhappiness, both for the overloaded person who cannot delegate because this false standard gets in the way, or for the delegatee whose performance is constantly criticized for reasons which are actually superfluous to the objective of the task.

Successful delegation involves straightforward principles:

- delegate everything which can be successfully handled by someone else, thereby freeing up time for things which cannot be handled by other members of the team
- choose the delegatee carefully, with knowledge that he or she has the ability, time and resources to carry out the task
- communicate the task clearly, and establish deadlines for completion; check that the delegatee understands, and will refer back before it is too late to correct or improve any shortfall
- step back and leave the delegatee free to get on with it
- give feedback and thanks to the delegatee once the task has been completed; this should lead to the level of work which can be delegated going up.

Key principles of time management

Determining goals and objectives

Design practice is about bringing new things into existence, and it is a prime case for determining the desired objective or result, and then looking backward to determine the steps required to reach this goal. Goals require definitions, methodology and deadlines, and these elements form the basis for all project planning.

All successful people are planners, but the plan alone is not sufficient. It must incorporate ongoing review and revision, and also contain its own cushions, safety-nets and back-up provisions. Once a plan has been established, it should be written down and communicated clearly to all members of the team.

Key result areas

This concept applies to both the consultancy as a whole, and to individual employees. It involves knowing what is wanted, and directing all efforts consciously towards the achievement of results in the areas which really matter.

For employees, the job description and performance appraisal functions described in chapter 18 will provide good indicators on key result areas. For consultancies, the brief, principal presentation points, and the company image and USP, determine the areas where results will be evaluated.

Establishing priorities

Time is finite, and decisions on its use have to be made on a basis of priority. The ability to prioritize and act accordingly is one of the key elements of advancement in any company, because it has such a fundamental influence on results.

Incoming propositions can be evaluated on the following scale:

A – essential
B – desirable
C – would be nice to do
D – delegate
E – eliminate

Eisenhower decision grid		
	Urgent tasks	**Non-urgent tasks**
important tasks	1 vital and urgent	2 vital and non-urgent
less important tasks	3 less than vital but urgent	4 only if time permits

In most organizations, 80 per cent of time is spent on category C, D, and E work, with the remaining 20 per cent having the greatest influence on results. It is important to learn to prioritize immediately, and not to accumulate issues which wait for consideration. Once priorities have been established, tasks should be tackled in order of importance, with all 'A' work being cleared before anything else gets attention. In establishing priorities, it is also necessary to consider timescale, which means evaluating both the importance and the urgency of the proposed task.

The chart above shows a simplified version of a decision grid originally introduced by Dwight Eisenhower. Assess tasks according to their urgency and importance and write them down in the appropriate box. Items which appear in box 1 merit maximum attention and resources; those in box 2 can be scheduled into the timetable and hopefully attended to before they become urgent. Box 3 comes next, but the person concerned must be sure that the task really does need doing; if time permits, items in box 4 can be considered.

The word 'no' is the greatest timesaver of all, and should be used to control all tasks which do not get 'A' or 'box 1' rating, and to eliminate irrelevant work.

Decisiveness

People who get a lot done do it by being decisive. Their guiding principle is 'do it now' and they avoid building up a backlog of tasks which they hope to get around to sometime, but don't know when.

Consider how much can be achieved before going away on holiday. The holiday departure date becomes a watershed, and tasks must be placed into 'before' and 'after' categories. The big push to get away often clears work from the bottom of the heap into the wastepaper basket, because it is finally recognized that there will never be time to do it.

Decisiveness involves passing things on quickly and not becoming an organizational bottleneck. Try to handle pieces of paper only once: read, decide and either implement or delegate.

Consider the following offering from the prolific writer, Anon: 'Yesterday is a cancelled cheque; tomorrow is a promissory note; today is ready cash.'

Basic time-effective working habits

Organizing the workspace

A clean, clear workspace is the best setting for creative thinking. It avoids time spent looking for lost items, and time lost trouble-shooting as a result of things being overlooked and fouling up. It is much easier to be organized once a simple storage system has been established; consider using rigid card magazine storage boxes which can stand together on a shelf, labelled cardboard boxes, hanging files, plan chests, whatever suits and will get used.

All material retained as 'current' should fit in to one of these categories:
- immediate action
- medium-term action
- pending
- reading material.

Things to be kept for reference should be filed on an ongoing basis, into easily accessible categories, otherwise they will take up space, but not be of any practical use. Anything which doesn't fit into any of the above pigeon-holes should be either passed on, or thrown away.

The discipline of clearing desk or table at the end of the day is a good one, and also provides a good starting point for planning and prioritizing the following day's work.

The three-file system

This is a simple approach to controlling paper which can pay dividends to even the most anti-organization person. The three files are:
- waiting for reply
- pending
- forthcoming events.

The file copy of any letter sent out goes into 'waiting for reply' until either (a) a reply arrives or (b) the sender decides to chase it up, in which case the chase letter is added to the file. Things which are not waiting for a reply, but not sufficiently resolved to go into a subject file for storage go into 'pending', and papers relating to any future commitment, such as agendas, tickets, conference details go into 'forthcoming events'. All files have to be reviewed regularly, and necessary action taken and noted.

Once papers leave the three files, they will move into subject or project files for long-term reference and storage; papers to be filed can usefully be marked 'file/(name of file)' in the top right-hand corner. The golden rule is never to file anything which is not completed.

The essential tool of time management

The single key to using time more effectively is the use of a diary, planner or log-book. There are various commercial time-management systems on the market, but there is nothing particularly magical about these products unless they succeed in persuading the owner to use them.

Selecting a system

The physical form of the planner can be a loose-leaf diary system (such as Filofax), large-format diary or a notebook pc; the most important thing is that the user is comfortable with the system selected and finds it both easy to use and carry around. If a diary or planner is to be used as an effective time-management tool, it must accompany its owner at all times. With effort and continued use, it will develop into an external brain, fulfilling the function of appointment calendar, address book, telephone directory, record of activity, planning tool, memory aid, expenses record, monitoring device, reference book, and catalogue of ideas.

The choice of system should take account of all these functions; a loose-leaf or pc-based system allows easy movement of data such as addresses and phone numbers from one year to the next, and also offers ready-made reference elements such as maps, and conversion tables. It is also essential to choose a system which provides sufficient space for entering a quantity of material, without becoming too bulky.

A lot of time is wasted by people not taking diaries to meetings, which prevents future arrangements being settled on the spot with all parties present.

Using a planner effectively

Within the planner, it will be possible to construct yearly, monthly, weekly and daily plans, and to review and refine these on an ongoing basis. The planner reinforces the principle of putting things in writing, and it should contain the following elements:

- detailed time schedules for each day
- sections of time blocked out in advance, for holidays, 'research and regeneration' time and 'back-up' time (see below)
- notes of all deadlines, and time allocated in advance for work to be carried out or necessary preparation
- notes of any commitments made, eg 'send James invitation for company party'; these should be made at the time of undertaking to do something, checked regularly, acted upon, and ticked off when completed.
- notes of phone calls to be made or received, on the appropriate day
- records of things which should happen on specific days, eg 'photographs for airport job to come from John C'
- details of forthcoming events such as lectures/exhibitions, for which time will be allocated according to other commitments
- notes of expenses
- full details of all meetings, to include name of person plus telephone number and full address of the venue, and so on
- notes of when colleagues/clients/suppliers etc. are on holiday or unavailable.

The planner should be reviewed regularly, and checked thoroughly at least once a week at a specific time, eg Friday afternoon. During this session, a check should be made that everything scheduled for the current week has happened or been acted upon, and, if necessary, things carried forward into the following week. At the same time, commitments for the week ahead can be reviewed and planned in more detail.

People who work closely together

When people work very closely together, such as managers and PAs, or team partners, their plans become interdependent, and it will be necessary to run two planner systems. In this case, one person should take overall charge of the system, but each should still carry an up-to-date copy at all times.

Regular diary-planning sessions are a great help. Material can be stored between sessions for consideration and decision at a pre-arranged time and the meeting provides a valuable point of contact and liaison on the workload of both parties.

The other essential technique for time management

This device is so simple it sounds silly, but is of great value. Make a commitment to empty everything out of bag, handbag or briefcase, once a week. Check what is there, and transfer it to where it should be.

Daily plans

Life happens one day at a time, and 'today' is the vehicle through which we take part in it. The day is the smallest and most manageable unit in time planning, and day plans are the means by which longer-term planning is both controlled and implemented. Using the planner, the first task of each day should be to draw up a daily schedule and a 'to do' list, taking account of the following points:

- the daily schedule contains only things which must be handled that day, and it should be both realistic and attainable
- start out by listing activities:
  - work commitments
  - planning/preparation
  - anything carried forward from previous day
  - telephone calls/letters
  - regular activities, eg staff meetings
  - appointments
  - social commitments
  - personal commitments
- some things will be time specific; others should have estimated times apportioned and then be given time slots in the schedule
- bunch together similar tasks, eg all outgoing telephone calls
- don't mix creative tasks with functional tasks
- develop blocks of time; a minimum of 60 minutes is needed for serious work (it takes 30 minutes to get into it); 3 hours uninterrupted work is the ideal
- don't attempt to work for too long without a break.

Once the daily schedule and 'to do' list have been assembled, stick to them. Don't do anything which isn't on the list, and take requests for action on board for inclusion in future lists, instead of allowing them to destroy the current one.

The day is not completed until the list has been finished, and reviewed. Mark off everything which has been done, and keep the list for future reference, eg 'how long did it take to do the concepts for x?'.

Once the habit of working to a daily schedule has been established, it will lead to better overall diary planning. Try to block out certain days as 'no appointments', perhaps immediately after a business trip, before a major presentation, or as a regular slot for serious research and creative work on a set day each week.

Controlling day-to-day reality

Time-management specialists talk about 'time thieves' which prevent people meeting their goals and objectives. There are many different time thieves in design consultancy, but the principal ones are:

- telephone calls
- interruptions (usually by colleagues)
- meetings
- crises.

All of these things are constant problems which threaten the best constructed plans and timetables, but there are a number of useful techniques which can be used to control their impact.

Telephone calls

Never answer a telephone call unless prepared to deal wholeheartedly with whatever is on the other end. This means arranging, if possible, to have certain periods of the day away from answering the phone (using a colleague or an answering machine), particularly if tackling a concentrated work period. Don't answer telephone calls on the way to an appointment; it is a sure-fire route to arriving late and/or forgetting to deal with the caller's business.

If calls are being screened, handle this diplomatically. The person answering should offer a selected excuse (eg 'she's out of the office at present') before asking for the caller's name; to do otherwise implies that the response depends on who is calling.

Planning phone calls is a great timesaver, and it is helpful to bunch together both making and receiving phone calls into a particular part of the day.

Sometimes an unwanted telephone call finds its way through the net, but if necessary, it is still possible to escape. An excuse such as 'I'm afraid I have a client waiting in reception, could I call you back later?' should give no offence and enable the call to be returned at a more convenient time.

Interruptions

External interruptions can be screened in the same way as telephone calls; it is important that the unwanted visitor should not establish direct contact with the person to be visited, otherwise the interruption will be less easy to control.

The major problem in design consultancies is internal interruptions, arising from open-plan studios and collective working. Some studios adopt a 'traffic light' system, where staff may display a red card if they don't want to be interrupted. In general, companies need to work on achieving sensitivity in internal contacts and the lead must come from the top (managers are often the worst offenders). With advance planning, the need to break into someone's time becomes less, and an interruption should always be prefaced by 'Could you spare me x minutes?' or 'Can you talk about the y project?'. The other side then has the option to suggest an alternative time.

People who are subject to constant unwanted interruptions might try:

- standing up, holding a piece of paper and claim to be just leaving
- saying 'I can't talk to you now - would you like me to ring you when I've finished?'
- in a particularly undisciplined studio, moving elsewhere (eg canteen, conference room, home) in order to achieve continuous work time
- scheduling sensitive work for quieter times of the day, such as early morning or late evening, and adjusting the standard working day accordingly.

Meetings

On average a manager spends two hours a day in meetings. If those meetings are not effectively run, only half of the time will be productive, resulting in an overall loss of six weeks' time per year, per manager. Management must set the tone for time-efficient meetings, and this should include the following:

- don't hold any meeting unless absolutely necessary; involve only those people who must be there, and limit their attendance to the relevant parts of the meeting
- prepare an agenda or objectives for the meeting in advance; circulate any papers or background material beforehand
- make sure everyone arrives on time and start the meeting without any latecomers
- choose an efficient leader for the meeting, who should outline the objectives and the timescale at the outset, also taking responsibility for keeping everyone to the point and rejecting unscheduled items
- ensure that there will be no interruptions
- summarize decisions at the end of the meeting, and arrange any follow-up date while everyone is present (with their diaries).

Crises

Crises are of two types; the ones which can be avoided or mitigated by contingency planning, and those which actually happen.

The latter group should be kept to an absolute minimum by good planning. Murphy's Laws are usually true:

- nothing is as simple as it seems
- everything takes longer than expected
- if anything can go wrong, it will.

Good plans must therefore be based on an average to bad scenario, rather than shining optimism. Planning should always include contingency time and money, and an anticipation of where hitches and problems may occur, with appropriate safety-nets built into the scheme. These can include the maintenance of good relations with back-up staff, freelancers, contractors and suppliers, all of whom can be invaluable in an emergency.

Crisis is both addictive and contagious; if, despite all precautions a full-scale crisis erupts, an effective counter strategy should be available. This might include the following pointers:

- act quickly and decisively
- put one person in overall charge
- don't overreact, and don't allow the panic to become an occupation in itself
- contain the crisis within its own area
- collect all possible solutions, together with any relevant back-up information
- evaluate possibilities and make decision(s)
- implement the solution, delegating as many tasks as possible
- tell everyone (including the client if relevant) what is going on and how the problem is to be resolved.

Summary

Skill in handling the use of time is one of the principal methods of improving both output and quality of performance in design. The most brilliant solution is unlikely to find favour with the client if crucial deadlines have been missed, and the ability to cover more ground, effectively, within finite time allocations is a powerful skill. In time mamagement, the designer addresses a balancing act, between pushing boundaries ever outwards, in search of the better solution, and at the same time meeting deadlines and harmonizing many different events and processes towards a common goal. The saving grace is that time management cannot only be learnt, it also improves with practice, and, consequently, the barriers of the possible can be pushed gradually further and further towards the extraordinary.

Further reading

Mackenzie, Alec (1985) *New Time Management Methods for You and Your Staff.* New York: McGraw-Hill.
Mackenzie, Alec (1991) *Time for Success.* Maidenhead: McGraw-Hill.
Seiwert, Lothar J (1989) *Managing Your Time.* London: Kogan Page.

Video learning packages

The Time Manager. BBC Training Videos.
*Working Effectively: Managing Time.*The Open College.

Audio learning package

The New Time Management. Nightingale Conaut.

Mathew Bell/Sharp Practice

'All things are difficult before they become easy.'

Persian proverb

9 INTERPERSONAL AND NEGOTIATION SKILLS

Shan Preddy

The art of dealing with other people is so much a part of life in a design consultancy that it is often taken for granted. Without other people – clients, colleagues or suppliers – no design project would ever reach completion. However, there are ways in which interpersonal skills, or dealing with other people, can be improved, so that work can be both more effective, and much smoother in its operation.

This chapter develops issues discussed in chapters 4 and 7. Specifically, it deals with negotiation skills; assertiveness and overcoming objections; and problem solving.

Negotiation

Skill in negotiation saves time, money, anxiety and frustration, and avoids discontentment with the outcome. To negotiate successfully is to exchange something of lower value (to one party) for something of higher value (to the other party). Negotiation is used when at least two parties control or influence the outcome. The desired outcome is shared by all parties, even though the individual objectives may be very different. If one party alone controls the outcome it is a command or a unilateral decision, and not negotiation. Equally, if the decision on the outcome is deferred to a third party, it is arbitration and not negotiation.

Negotiation is usually associated with financial transactions. In practice, however, negotiation takes place every time one person tries to get someone else to agree to or do something. The principles and practices involved apply whether the negotiation is over fees, contracts, employee benefits, company policy, approval of design work, time schedules, when to go on holiday, or where to go for lunch.

There are three things which make a good negotiator: knowledge, skill and attitude. Some would say that the greatest of these three qualities is attitude. Often, the most significant element in the likely success or otherwise of any negotiation is the attitude of either party towards the negotiation in question, and towards their chance

Qualities of a good negotiator

Knowledge of
- negotiating principles and strategies
- the context of the particular negotiation
- the detailed subject matter involved
- the negotiator's own strengths and weaknesses in general, and in the particular situation.

Skill in
- the analysis of the issues involved
- personal interaction, persuasion and motivation
- communicating what is wanted
- listening, and being patient.

Attitude towards
- negotiating in general
- the specifics of each negotiation
- the negotiator's will to succeed
- creating a satisfactory outcome for both parties; giving and getting, compromise or concessions.

of success. Any negotiating position therefore depends on two factors; reality, and the perception of one party by the other.

Sources of power in negotiation

If control of the negotiating position is conceded, it becomes difficult to achieve the desired objective. Control is only relative power. All power is relative; absolute power is only relatively absolute. Power is determined by what we think of the other party, and of ourselves, and there are several useful sources of relative power.

Negotiating positions

The seller, whether of goods, services or ideas, tries to ascertain the maximum that the buyer is willing to pay, without disclosing the minimum that the seller is willing to accept. The buyer, on the other hand, tries to ascertain the minimum that the seller is willing to accept, without disclosing the maximum that the buyer is willing to pay.

Before entering into any negotiation (whether financial or non-financial), it is important to identify the parameters which will influence the negotiating position:

- what is the best possible deal?
- what is the acceptable minimum (or maximum) level of finance or other arrangement, below (or above) which the deal will not be made?
- what concessions can be made, if necessary?

Expectations need to be high. Reaching for the stars means getting to the treetops easily. Reaching for the treetops means the likelihood that the project will be lucky to get off the ground. Starting with a shockingly high (or low) but credible opening offer results in the other party's belief in their own price or position being shaken. Remember that professional buyers are trained to 'squeeze until the pips squeak'.

Ten sources of power in negotiation

1. Knowledge: about the subject matter
2. Skill: in negotiating techniques
3. Attitude: confidence in success
4. Competition: the ability (whether real or perceived) to go elsewhere
5. Legitimacy: proven right
6. Persistence: standing ground
7. Withholding information: determining what information will be given to the other party
8. Withholding opinion: telling a supplier that they are the perfect or the only choice for the job reduces their incentive to negotiate
9. Patience: waiting, listening, learning; wherever possible, letting the other party open the negotiation first so that knowledge can be increased
10. Silence: the power of silence should not be underestimated; it can cause the other party to panic and make concessions

Concessions

Successful negotiation is not about winning or losing; it is about achieving results. This can often require compromise. The phrase a 'win/win' attitude or situation does not refer to a desire to win at any cost. It refers to an outcome which is satisfactory to both parties. In other words, both parties 'win'. This is possible because both parties usually want from the negotiation something slightly different; each may have ideally wanted more, but each is satisfied. It is not a question of altruism; if both parties are happy with the outcome, each will work hard to make the agreement succeed. Conversely, if one party is unhappy with the agreement, they will do their best to undermine it. A win/win result is only possible if both parties are clear about their own objectives, and are willing to make concessions. The aim is to make smaller concessions than the other party, the 'middle ground' never really being in the middle. Flexibility in approach helps, with changes made as the situation demands. Sticking rigidly to a position will not encourage a shared outcome.

Preparing for negotiation

What are the issues?

What are the objectives (of both parties)?

What tactics are likely to be used by the other party?

What facts or information support the case?
- current market rates
- inside knowledge of the other party's needs (eg are they about to go broke? Is their timing schedule very tight?)
- competitive rates (both parties)

What facts are the other party likely to have available?

What external factors are likely to affect the outcome?
- market economy/climate
- supply and demand
- time constraints
- legal implications and considerations
- past precedents, standard practices.

The negotation meeting

For any negotiation, preparation pays dividends. It is often useful to find out, or work out, as much as possible about the situation in advance. Meetings involving negotiation usually follow a pattern.

1 The warm-up

This should be friendly but businesslike. The time can be used to assess the other party's mood, and state of mind.

2 Statement of joint desired outcome

It is helpful at this point to state the shared desired outcome. Ideally, real objectives should not be disclosed; rather, the desired result should be stated.

3 Declaration of opening positions

If possible, get the other party to start first; listen and learn.

4 Disagreement and conflict

There needs to be a full exploration and discussion of any differences between the parties. Positions should be probed, questioned, challenged, rigorously tested and weakened where possible. However, questions are more productive than demands, which can be threatening. (See also the later section on questioning techniques.)

5 Offers

Concessions, adjustments, and compromises are stated; alternatives are explored. It is productive to constantly repeat and reaffirm the agreed points.

6 Agreement

The agreement or settlement is summarized in principle ('subject to . . .'). Everyone must understand the same thing; if issues are fudged, problems will occur later. It is always helpful to confirm in writing the agreement made at the meeting as soon as possible afterwards. The party which does this gains a psychological advantage over the other party, by using a strong confirmation as an opportunity to take control of the situation. This is a form of power display.

Strategies and dirty tricks

The difference between a strategy and a dirty trick largely depends on which party is using it, although some strategies are more ethical than others. The decision on whether they are used or not is personal. However, dirty tricks should be recognized, and dealt with if used. A few basic strategies follow. (For more details of strategies and examples of their use, see Fowler 1990 and Kennedy 1991.)

Salami strategy

If agreement to the whole thing cannot be obtained, agreement to certain elements might be possible. In other words, if someone won't give away a whole salami, they may be persuaded to give away a series of slices, until the whole salami has gone. Once people start saying 'yes', they are more inclined to continue saying 'yes'.

Design companies already use this technique by quoting for stages of a project, rather than the whole commission. The same approach can be applied to other areas, such as getting difficult creative work approved.

The two-letter word strategy

The most useful word in a successful negotiator's vocabulary is not 'no', but 'if . . .'. It leads to concessions from the other party. For example:

- 'If you drop the price on this to £x, I can arrange to authorize the order for you.'
- 'If you give us a credit on this annual report and allow us an additional 200 copies to use as promotional material, I will personally attend the print run.'
- 'If you will guarantee payment in advance for each of the design stages, we will be able to reduce the price slightly.'

Don't change the price, change the package

Prices are challenged for a variety of reasons:

- a love of negotiating
- a genuine lack of available budget
- an expectation that the price has been set at a level which allows reduction (eg house prices, antiques)
- competitors offering a cheaper price
- an excuse for rejection; it is easier to say the price is too high than to say that the work or the company or the individual is not liked or is below standard.

If a price challenge occurs, the price should not under any circumstances be lowered. The price quoted is for the work outlined. A different price would mean different work; different work would mean a different price.

The package should not, however, be increased for the original price. A retail outlet can use this tactic successfully by offering free additional goods. A design company cannot do this. The client's reaction would be: 'If you can afford to give away all of these extra things, you can afford to reduce your quote . . .'. If work is done for a client cheaply, or free of charge, a precedent has been set, and a weakness demonstrated. It does not soften the other party up; it toughens them. An example of this is the way some design buyers now expect free creative work to be supplied as a part of a pitch. Historically, the design business has 'trained' them to do it.

Generosity in negotiation is not contagious. If sprats are thrown away in the hope of catching a mackerel, the only guarantee is that the sprats will be lost. Moreover, if a price is lowered even once, it will be difficult to enforce the full amount in the future with the buyer in question. As a result, there will be a temptation to artificially inflate future estimates, which will then seem expensive in relation to the competition.

Feel the quality

As design companies do not operate in a fixed-price market, cost estimates will vary hugely. Consequently, it is important that a significant marketing point of difference for the consultancy is developed, so that the apparent competition is reduced. Why should a client appoint one design company instead of another? Staff must be able to summarize the benefit of using the consultancy in question to any price challenger.

Test the strategy

To test the strength of a position relative to that of the other party, a series of questions about the outcome can be asked: 'If we agree this, what if . . ?'.

This 'what if . . ?' strategy is particularly useful if there is a suspicion that the other party has not revealed everything, either by omission or commission. It is also very useful when checking through a contract, whether it has been specifically drawn up or standard; gaps will often reveal themselves in response to the 'what if . . ?' question.

Deadlock or 'I'll see you in court'

If someone refuses to negotiate, it could be a deliberate ploy. Begging them to negotiate puts them in a very strong position. A discussion about why they will not negotiate should be instigated, either directly or through a third party. Letters can succeed here, because they remove emotion. However, the written word is a permanent record, and a trusted colleague or solicitor should examine the letter before it is sent. Sometimes, a third party trusted by both parties can be an effective intermediary. Deferring a decision to another party or a higher authority (arbitration or judgement) is also possible. The risk is that the outcome might not be satisfactory; previously determined minimum/maximum concession parameters might not be met.

Assertion and overcoming objections

Assertion is not the same as aggression. Assertion comes from a position of confidence, or strength; aggression comes from a position of defence, or weakness. The ability to be assertive when necessary is invaluable.

One use for assertion is the ability to say 'no' to a request in such a way that it almost sounds like 'yes'. In other words, it allows a refusal to be made in a positive manner, by persuading the person who is applying pressure to apply it elsewhere.

Examples of this approach would include phrases like:

- 'I would love to do that for you but, if I do, I will have to stop working on your other project. Which would you like me to do?'
- 'I would be pleased to help, but I can't do it right now. I will do it for you next Thursday, if it's any help to you.'
- 'It's really important that we discuss this, but I haven't got the time to do it justice at the moment. Can we fix another time to discuss it properly?'
- 'Wouldn't Jane be better at this? She is really talented at this kind of work.'

Calm, cool, collected

Assertion is also an important skill for overcoming objections. The following CAT mnemonic might prove useful:

c – cool, calm, collected
a – appreciation
t – turn on the broken record

By appearing to be cool and calm, a further source of power can be harnessed. Once temper is lost, assertion will soon become aggression, and the high ground will have been deserted. Assertion gains respect; aggression loses it.

Appreciation

Demonstrating an appreciation of the other party's point of view will dilute any aggression on their part and slacken their firm resolve; it is difficult to be fierce with someone who is apparently in agreement. Of course, the two parties are not in agreement. One is simply stating that they understand and appreciate the other's case.

Turn on the broken record

The constant repetition of a position demonstrates commitment to it. The words used may change, but the position remains unaltered. Skilled broken record operators include politicians and children; much can be learned from them.

Questioning techniques

One successful way of demonstrating assertion rather than aggression is to learn several questioning techniques; questions are always less threatening than statements or demands. Careful and accurate listening is one of the hardest aspects of good communication, and ability in the use of questioning techniques helps. Questions also help to overcome objections, particularly if those objections are irrational or emotional. The constant asking of detailed questions in order to define the precise nature of an objection will at worst clarify the issue, and at best resolve it.

In addition, questioning techniques can be useful when dealing with criticism from others, for example in difficult meetings, in appraisals, and in counselling.

Open questions

A closed question encourages a 'yes' or 'no' response, or a short one-word answer. Open questions demonstrate an interest in the other person's point of view, show that their opinions are valued, and allow a better understanding of their needs. Asking people to express their feelings or opinions can lead to a fruitful discussion. The six question words are helpful: what/which? why? when? how? where? who? An example might be: 'Why do you think your boomerang won't come back?'

Reflecting questions

In this technique, the statement the other person has just made is repeated in the form of a question which selects the most important idea, feeling or concern stated.

Reflecting questions avoid arguments; there is no need to agree or disagree with what the other person has said. In addition, a reflecting question will confirm the hearer's understanding of what has been said, which is particularly useful if dealing with an inarticulate person. It will also demonstrate active listening. A reflecting question should be followed by a further open question. For example: 'Are you saying that your boomerang is defective in some way? Why do you think that?'

Directive questions

A reflecting question is likely to get a 'yes' response, so there may be a need to probe further. Directive questions can be used to continue a dialogue, or to explore a specific issue. For example: 'Your boomerang looks alright to me. If you took a few lessons in boomerang throwing, mightn't you be more successful?'

Conflicts and problem solving

Problems happen. If the problem is accompanied by time pressure, it becomes a crisis. The worst person to solve a problem is the individual in the centre of the crisis, if he or she is panicking.

The following method of finding solutions is very effective. It is particularly useful for longer-term strategic thinking, but, if learnt and assimilated to be as natural as breathing, it can also be used in a crisis. The method has three steps.

1 Turn the problem into an objective or goal

Most problems are expressed in the negative. 'I can't . . .' or 'There isn't . . .' or 'They won't . . .'. The first step towards defining an objective or goal is to express the problem in the positive. 'I would like to be able to . . .' or 'There should be . . .' or 'We want them to . . .'. That positive expression then becomes the objective.

2 Generate options

List the actions that are under personal control which could positively influence the outcome. The options listed should be very specific, giving details of what is to be done, how, by whom, where and when. Although this method of problem solving is linear, an element of lateral thinking helps here. For strategic planning, the involvement of several people in a synectics (or brainstorming) session can be a fertile source of options. For crisis management, however, one individual's options are generally preferable.

3 Test the options, each in turn

First, the specific actions identified in the options should lead directly to the defined objective. The 'what if . . ?' strategy mentioned above can be used here. Second, the option must lead to the correct objective. If there remains any doubt that the objective, once achieved, is satisfactory, the objective should be revised.

Only when the best possible option leads to the achievement of the best possible objective, should the process be considered complete. It is sometimes necessary to revise the options or the objective repeatedly; however, with practice, accurate objectives can be created very rapidly.

Conclusions

Life would be very simple without other people to complicate things, but it would also be boring. Business would be just as simple, and just as boring, and, in addition, much less effective. As outlined in chapter 7, the sole purpose of management is to achieve results through other people.

Once the theories of dealing with and negotiating with other people are learnt and used, they will become a natural part of daily business life. Practice may not make perfect, but it can make permanent.

Further reading

Dyer, Wayne W (1988) *Pull Your Own Strings*. London: Arrow.
Fisher, Roger and Ury, William (1989) *Getting to Yes*. London: Business Books.
Fowler, Alan (1990) *Negotiation: Skills and Strategies*. London: Institute of Personnel Management.
Karrass, Gary (1987) *Negotiate to Close*. London: Fontana.
Kennedy, Gavin (1991) *Everything is Negotiable*. London: Arrow.
Lindenfeld, Gael (1987) *Assert Yourself*. London: Thorsons.

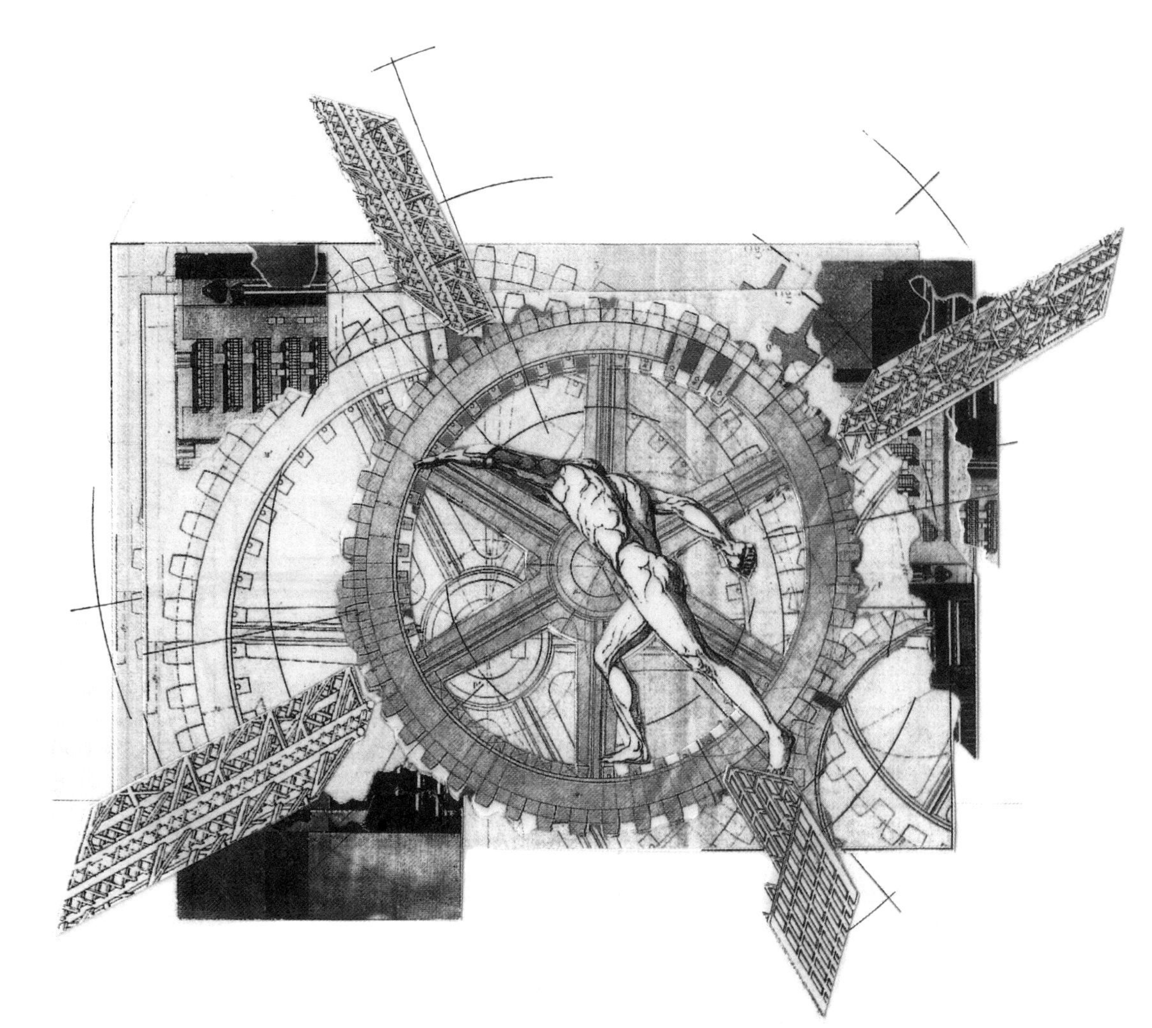

David Loftus/Sharp Practice

'Marketing is human activity directed at satisfying needs and wants through exchange processes.'

P. Kotler

Marketing Management: Analysis, Planning and Control

10 MARKETING DESIGN CONSULTANCY

Liz Lydiate

Design consultancy specializes in providing creative solutions to the marketing needs of clients, but frequently finds it much more difficult to apply these skills to the marketing of design generally and/or the work of an individual consultancy. Part of the problem is designers' commitment to their work: they tend to have a strong personal belief in the importance of design, and, although this is a great strength it also gets in the way of objective judgement and pragmatism in tackling activities such as marketing. In addition, the design industry is addressing a collective audience which (in general) finds itself competent to denounce bad design, but (also in general) largely unconcerned with the pursuit of good design practice.

These are the difficult parameters within which the marketing of design consultancy takes place; at its bleakest, the design industry can be seen as offering an outstandingly versatile and effective product to a market that frequently doesn't understand the product on offer, or how to make use of it. In addition, of its very nature, design consultancy is at the outset invisible. The marketing activity must succeed in persuading the client to embark on a partnership of professional trust which is in itself the only route to bringing about the desired result.

Responsibility for marketing

The activity of making creative work requires a personality which will look within itself for inspiration, execution and the setting of standards. This is essentially a solitary and private process, and often sits uneasily with the other necessary part of the working life of a designer – marketing the work.

Many design companies now employ specialist new business and marketing staff, and this can lead to designers thinking of this part of the work as none of their concern. Marketing is part of the responsibility of every person working in a design company, and an overall understanding of marketing principles in relation to design, together with a detailed understanding of how any one job keys into the overall marketing strategy of the company, is essential for success. This premise derives from the nature of design consultancy. It is a creative service, where many different skills and talents are brought together to achieve a result for the client. The client is buying the eventual design solution, but must also buy, and experience, the process.

Because the marketing function involves not only getting work, but also carrying it out, delivering the result (and hopefully, achieving repeat business), every single member of the team can, and does, influence marketing performance. It is the interface between the client's requirements and the creative process which produces the design result. Even if new business people and account handlers are involved, this should not result in designers having no direct contact with the client. The designer is an advocate for the work, and is often the best person to communicate ideas to the client and build confidence in the more adventurous solution.

Even though it is not seen as such, the creative work itself is also part of the marketing process. It is the client's direct experience of the mysterious and invisible commodity which she or he is buying. The best design work delivers, or even improves on, the original marketing offer.

Design: marketing a service

Marketing a service differs fundamentally from marketing a product, which can be picked up, tried on, test-driven or otherwise checked out before buying. Services are intangible and invisible, and must therefore be bought on the basis of the reputation, track-record and known expertise of the company or organization. Design is a more intangible service than most, in that every design project is unique and the exact nature of the end result will not be known until long after the buying decision has been made. Contrast this with the buying of other services, such as air travel or dentistry, where the purchased outcome is much more specific, and the scale of the challenge in marketing design becomes apparent.

The situation has been further complicated by a traditional emphasis within the design community on the importance of the end result rather than the process and associated quality of service. In order to market a service effectively, quality of service must be placed on an equal footing with quality of product. This requires a degree of adjustment of attitude and values within some sections of the design industry as currently established. In addition, many clients do not understand the nature and true function of design; and even if they think they do, they consider it quite reasonable to expect to 'see before buying'.

The arguments against the provision of unpaid, speculative work form yet another layer of complexity in the marketing of design. Most design companies accept that creative solutions are based on extensive research and study of client requirements and a fund of ability, knowledge and inventiveness, which is, in effect, the consultancy's stock in trade; thus it is inappropriate to provide these services without payment. It is necessary and important to convey an understanding of these issues to potential and actual clients as part of the marketing activity.

Understanding market positioning

'Market positioning' is the term used to describe the way in which a company places its offer within the market as a whole, and how this decision will inform the resulting marketing strategy. As an example, a retailer will place stores in locations which reflect the nature of the company offer, and will have them designed in a style which also is in keeping. Stock selection and pricing policy will follow the same alignment.

A design consultancy will determine its market positioning in relation to its:

- skills and expertise
- experience in particular size fields
- aspirations and/or corporate plan
- culture/character
- resources

- opportunities
- competition
- environment.

An examination of market positioning should include aspirations (what the company would like to be) and perceptions (what it is seen to be). Sometimes these differ.

Factors affecting market positioning

The SWOT analysis technique described in chapters 5, 7 and 14 is the foundation for examining and understanding a company's market positioning, and the following notes amplify this stage in the establishment of marketing strategy. In addition, each point can be broken down individually into another SWOT, and the results reassembled to give a detailed overall picture of the company and its offer.

- ***The product:*** or in relation to design consultancy, the service. What exactly is the company offering? This is not as obvious as it sounds. For example, marketing tutors point out that people don't want drill bits; what they actually want is holes. Similarly, clients don't want design consultancy, they want design solutions.
- ***The clients:*** who are they, where are they, and how many are there of them? Is the client constituency growing or shrinking? What factors are affecting them currently and in the future?
- ***Client needs:*** what are they? How can they be met by design? Are client needs likely to change? If so, how and why?
- ***Client perceptions:*** of design generally, design in relation to the client company's own activity, and of the design consultancy making the offer. Is the consultancy seen as it wishes to be seen? Does it know enough about how it is seen? Is the company perceived as having a USP (unique selling proposition), and what is it?
- ***Market segments:*** markets divide up into sections, or segments, each of which has a particular character or requirements. Which market segment does the design consultancy itself occupy? Which client market segments is it addressing?
- ***Competition:*** which companies represent competition within the same market segment, and which present a threat from adjacent ones? How does the consultancy measure up against the competition?
- ***Area of operation:*** related to market segments, but a separate issue, eg a consultancy may address food and drink as a specialism within fmcg (this is its client market segment) but operate geographically in the UK, continental Europe and North America. Alternatively, another consultancy may address a range of specialisms, such as literature, corporate identity and retail, but choose to restrict its activities to a particular locality, such as Scotland.

Once these critical factors affecting market positioning have been addressed by management, they should also be communicated within the company. Managers could usefully consider whether their staff could answer the questions above, and the knock-on effect of their level of knowledge on how they carry out their work and act as conscious or unconscious marketing ambassadors for the company.

Market research for design

When researching markets for their own work, design consultancies must select the most appropriate sectors in which to go looking for new clients from a vast range of possibilities. There is a great temptation to follow known or attractive routes, going after client companies who are visibly active as commissioners of design work, or who will clearly have reason to undertake a major project. Both of these approaches are perfectly sound, but will involve competing in a crowded arena. WH Smith, for example, reckons to get three unsolicited approaches from designers every day.

There is still a very large market comprising companies who do not yet use design either at all, or to the full, and design consultancies have much to gain by researching and contacting clients in this area.

Researching a prospective market

In looking at new markets for design, research usually divides into two principal activities: desk research and field research, normally carried out in that order, with resulting data fed into the development of the marketing strategy.

Desk research involves using data which is currently available to learn more about the prospective market. (See chapter 17 for information on sources and techniques.) It should find preliminary answers to questions such as:

- size of the market
- whether it is expanding or contracting
- demographic qualities of the market
- factors influencing the market (internal and external)
- character/culture of the market
- opinion formers/communication channels particular to the market
- extent of market use for design services
- nature of market need for design services
- design consultancies already active within the sector.

Field research is likely to involve visits, interviews, surveys and other targeted information-gathering activities, and will probably be focused on developing and defining material and issues identified at the desk research stage.

Approaching a new market

Whether addressing existing or new markets for design, success is likely to come from a starting point of knowledge. This knowledge is two-sided and comprises:

- the design company's familiarity with trade practice, issues and needs in the target market
- the prospective client's familiarity with design issues generally, and with the approaching consultancy in particular.

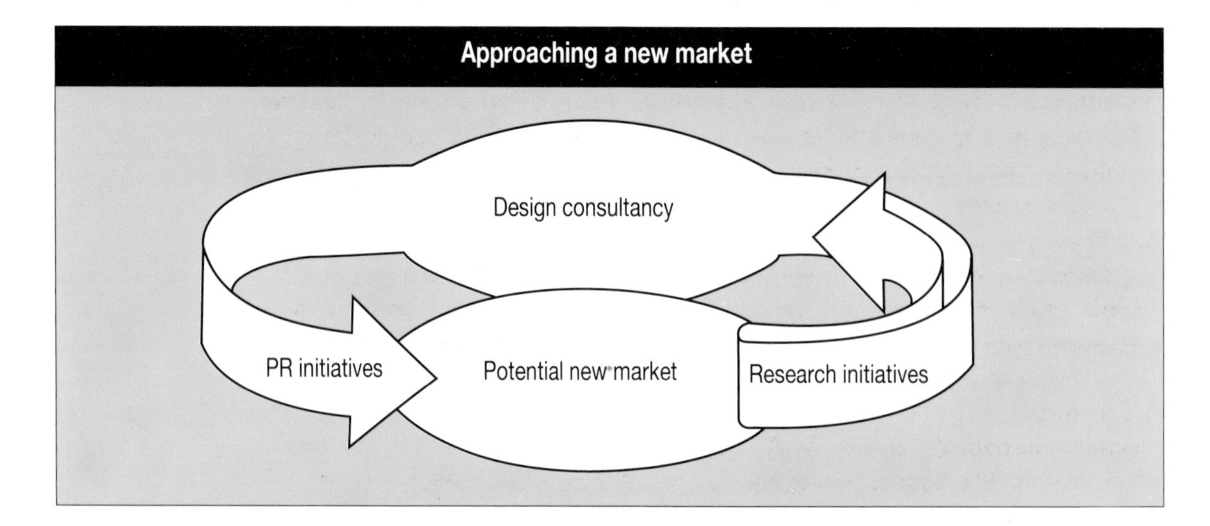

Achievement of the objectives in the first is most likely to come from market research and related planning and development activity, and the second from PR and communications activity initiated by design consultancies and design organizations as part of overall marketing strategy (see guidelines below).

Building a framework for marketing activity

Marketing the services of a design consultancy is likely to involve addressing both existing (core) markets and new markets. The method through which the marketing activity is planned has five principal elements:

- establishing the offer
- establishing the objectives
- establishing the message
- selecting the resources
- agreeing the plan and allocating responsibility.

The offer

Offer and objective are mutually dependent, and form the ground floor of the marketing framework. In a service industry, such as design consultancy, virtually all aspects of the company's activity form a part of the offer. They are likely to include:

- strategic planning
- knowledge of market/market needs, including specific sectors
- range and quality of skills within company
- methods of maintaining creative edge/competitive edge
- understanding the client's competition and appropriate differentiation
- knowledge and understanding of the client offer
- quality of service/project management
- understanding client culture and management style.

The objectives

Some companies subscribe to very vague but demanding objectives, such as 'to produce superb creative work' or 'to be the best design consultancy in Europe'. These goals are difficult enough to achieve in themselves, but are made more so if there is no clear and specific route towards achieving them. Company objectives should be defined in short, medium and long term, with each stage translated into straightforward, quantifiable steps, which can later be reviewed and analysed.

The message

Chapter 16 describes the increasing volume of information circulating in contemporary society. Many design consultancies add to this confusion by defining and communicating their offer in great detail using complicated language. Every consultancy should be able to reduce the core themes of its offer to a few simple phrases; this message should be known, understood and believed by all staff.

Selecting resources

In order to carry out marketing activity, the consultancy will select from a range of possible resources, and construct an overall plan. This reflects the process carried out by client companies; elements from which the company might select could include:

- print; brochures, leaflets, reports
- advertising
- inclusion in trade directories, databases, etc.
- direct and indirect approaches by phone or mail to target companies

- participation in exhibitions and trade events
- submissions for awards and competitions
- networking/membership of trade and professional organizations
- PR events/PR communications activities
- research into specialist fields and publication of results
- appearance of buildings/attitude of staff.

The actual selection made will be influenced by objectives, target markets, cost and available staff resources, but the effectiveness of the programme will be fundamentally influenced by its positioning within the working life of the company.

Allocating responsibility

Who actually carries out marketing activity varies from company to company, but the one person who will certainly be involved is the managing director or chairperson/chief executive. Beyond this, as with all management activity, clear allocation of responsibility and establishment of lines of communication are essential.

In some companies, PR and communications activity is regarded as separate from marketing, which is both wrong and counter-productive. The situation where designers are 'too busy' to provide information for PR and marketing initiatives is indicative of a company which no longer understands the business which it is in.

Agreeing the plan

Marketing needs support and commitment from within the company as a whole in order to be successful, and formal discussion and agreement of a marketing plan is an aid to clarity and focus. It can also form the basis for monitoring, performance evaluation, and knowledgeable future planning based on actual experience. The marketing plan should cover at least 12 months ahead, and be updated every month on a rolling basis. The plan should include:

- company objectives (or mission statement)
- detailed marketing objectives
- marketing budget, apportioned to activities
- full details of how the marketing campaign is to be implemented.

Issues in marketing implementation

Marketing is an enormous subject, and this is necessarily only a short introduction looking very briefly at four key issues relevant to the marketing of design consultancy.

Knowledge and choice: providing information

From an insider viewpoint, it is difficult to imagine anyone not knowing about the design industry. However, to the greater part of the world, design is shadowy and mysterious territory. Designers can benefit from asking absolutely basic questions about what information and knowledge clients have about design and design consultancies, where that information comes from and how it influences behaviour and decisions. In order to make use of something, it is necessary to know that it exists, what it is for and how to get some. Perhaps there will even be a choice, which introduces the need to compare, contrast and evaluate.

Impressions and decisions: influencing opinion

Decisions are not always made logically. A small but persuasive point can score a hit on the buyer's sensibilities, and will not subsequently be dislodged by new information. When design consultancies put forward information and presentations to clients there are certain factors that are likely to have major influence:

- does the client feel comfortable with the consultancy's people/attitude?
- is the consultancy able to offer detailed knowledge of the client's area of business? The ability to talk confidently and knowledgeably about trade practice/ developments/influences/areas of concern carry immense weight in new business approaches and must be followed through in subsequent performance.
- will the client recognize certain work included in the presentation? Familiarity increases confidence, and good exposure of work, achieved through PR coverage in media which reaches clients, has great value in laying foundations for recognition and understanding. 'Oh, you did that . . ?' is usually a good sign in a presentation.
- does the consultancy have a relevant track record and 'pedigree' with which the client feels confident? Prizes and awards are helpful in this area, but so too is working for other key players in the sector. The 'people like us' reaction is very important, and will relate back to the consultancy's understanding of its own offer and market positioning. Clients like to feel that they are using a consultancy which is known and respected amongst their own colleagues, and this confidence also provides a good base for future creative work.
- can the consultancy offer key, persuasive benefits at an early stage in the new business process? Children demonstrate this tendency very clearly when shopping, insisting on 'the trousers with the zip pockets', or the shop with the nice assistant. If correctly targeted, key benefits can take and occupy the high ground, and beat off challenges from elsewhere, as described earlier.

The design difference: value-added in client relationships

One of the strongest influential benefits is the potential relationship with the design consultancy itself, which, if correctly handled can have a stimulating and lasting effect on life and attitudes in the client company. Let's assume that both the design solution and the quality of service are excellent; this creative value-added can achieve both client loyalty, increased business and expanded volume of work. If a consultancy can be generous with insights into the creative process, new ideas or reactions to the client's business and an open door into a world of visual experiences, it can lift client relationships to new heights. Chapter 14 emphasizes the necessity of fun in the running of a design consultancy; this is about extending that fun and imagination to the client as well, and welcoming him or her into the company culture.

The case for 100-per-cent commitment

If a design company is well-run, all staff will share an understanding of and commitment to the corporate objectives and strategy. This of necessity includes marketing, and every member of the team can make a positive contribution.The results of successful marketing are a happy and prosperous company where everyone is doing what they like best, providing creative design solutions for clients.

Further reading

Cannon, Tom (1988) *Basic Marketing: Principles and Practice.* London: Cassell.

Hart, Norman A and Stapleton, John (eds) (1987) *Glossary of Marketing Terms.* London: Heinemann.

Lorenz, Christopher (1990) *The Design Dimension.* Oxford: Basil Blackwell.

Patten, Dave (1988) *Successful Marketing for the Small Business.* London: Kogan Page.

Wilmshurst, John (1984) *The Fundamentals and Practice of Marketing.* London: Heinemann.

*'Give me a firm place to stand on,
and I will move the world.'*

Archimedes

11 DESIGN IN BUSINESS STRATEGY

Shan Preddy

This chapter (which relates to chapter 2) explores the role of design in business strategy, and aims to demonstrate the ways in which design can be used as an effective and efficient business tool. Specifically, it contains:

- a review of how design is used in business
- design as cost or investment
- design as a strategic weapon
- current design management practices in UK companies
- what designers can do to help themselves.

Design in business

Design has an impact on every aspect of a business. It can create or improve the products or the services offered by a company, the environment in which the company operates, and the way in which that company communicates its benefits to its target market. Design, as used by business, falls into three broad disciplines:

- ***product design:*** a very wide field which includes all the goods manufactured by a company from trains to trainers, from paper mills to paper clips, from cars to carphones and from street furniture to tables and chairs
- ***environmental design:*** which covers the functional operation of a company; offices, factories, warehousing, restaurants, banks, retail outlets
- ***communication design:*** which deals with the way in which a company tells its target market about the benefits of that company, or its products or its services. This area includes corporate identity; product and corporate literature, including the increasingly important annual reports; packaging; brand identity; brand development; and events and exhibitions.

As a discipline, product design is closely allied to engineering and technology; environmental design is related to architecture; and communication design shares characteristics with other communication businesses such as advertising, direct marketing and public relations. These three disciplines, which not only require different skills and qualifications of the relevant designers, and often have different types and levels of seniority of client to work with, are quite different businesses.

However, they share certain pressures which are brought to bear on both manufacturers and suppliers of services. Pressure can come from within an organization; cost reduction and return on investment factors, production efficiencies, life-cycle expectations, and export requirements all affect the way in which design is commissioned and used. External influences also exist, often very powerfully.

The first of these pressures (see page 107) comes from the trade, or retail sector, which can influence the design of a product, an environment or a piece of communication by setting parameters on distribution, or warehousing, or simply by refusing to list, or stock, goods which do not meet their approval. Marks and Spencer's rigid quality control systems directly influence manufacturing practices.

The second external influence comes from governments, whether in the home country or in export markets, via advisory bodies or legislation. In packaging, legislation on use of materials in different European countries can mean variation in, say, plastic bottles used for Sweden or Switzerland.

The third external influence comes from the consumer, often driven by pressure groups. The strongest of these at present is the environmental lobby, which, coupled with the 1990s' reaction against conspicuous consumption, is having an effect on business decisions regarding design. Although consumer studies always show that consumer behaviour lags some way behind consumer attitudes, the growth in eco-friendly products and their presentation would suggest a permanent pressure in this area. Business results endorse the power of the guilt-assuaging ability of The Body Shop's offer; skin can be lavished with cocoa butter, hair smothered in fruit-based shampoos, baths scented with herbs, and the world is still being saved.

As well as being linked by these pressures, all three design disciplines are united by one further, essential factor: they are all able to provide a client company with a powerful business asset.

Design: cost or investment?

Money spent on commercial design should be an investment. As far as the commissioning client is concerned, the purpose of design is to create or improve the quality of new or existing environments, communications or products, and thereby improve business performance and, ultimately, business profitability.

If there is no improvement in performance as a result of the expenditure, then that expenditure has become a cost rather than an investment. The nature of the return on investment will vary from company to company, and from project to project, as will the timescale involved in seeing that return.

In Europe and the USA, companies are normally geared to expect short-term results, in anything between two and five years. Public companies are often expected to deliver results more quickly than those in private ownership. In Japan, on the other hand, the timescale for return on investment is more likely to be between ten and fifteen years. The Japanese attitude to expenditure on design, particularly on product design, is very different from that of western organizations.

During the 1980s, design became widely perceived as a significant corporate resource. Every chief executive in the UK, it seemed, was prepared to declare the importance of design to his or her organization. Government grants became available through the Department of Trade and Industry's Enterprise Initiative for design projects, and talk of the power of corporate identity was everywhere, helped by the high incidence of corporate mergers and management buyouts.

However, there is still a wide gulf between the image and the reality of design practice in business. Design accounts for a relatively small proportion of business expenditure. A regular, annual company budget of £5 million for any kind of design activity is rare. In contrast, a regular, annual advertising budget of £5 million for just one brand or product is not unusual. Does this mean that client companies are not recognizing the value of design compared with that of advertising?

A research survey carried out by the Design Research Company in 1991 reports that 71 per cent of clients do not have a set budget for design, but buy it on an ad hoc basis, despite the fact that most respondents strongly believed that they have a long-term marketing strategy. Does this mean that client companies do not, on the whole, view design as a strategic tool?

Design as a strategic tool

The widespread development of workable design management practices in British organizations has been slow; only a handful of companies recognize the real value of design, and have learnt how to manage it. Products are restyled when a complete redesign would be preferable, or they are designed according to existing machine capabilities. The importance of environment to sales performance has been recognized by parts of the retail sector, and by the more upmarket leisure and catering organizations, but many UK offices and factories are inadequately presented and equipped compared with northern European or American counterparts. Corporate identity is still perceived, despite the valuable work done by Wolff Olins, as the creation of a graphic logotype to be applied to stationery. Packaging, one of the most visible expressions of a brand's values, is often delegated to the most junior brand manager, with minimum budgets.

This differs significantly from some of Britain's competitors; companies such as Ford, Philips, Sony, Coca Cola and Swatch Watches, all have widely reported and well-documented approaches to design management, and a demonstrable commitment to the power of design. For example, Ford discovered in the mid-1980s that it was steadily losing market share to European and Japanese imports and to General Motors. It made a courageous leap in design terms towards a smaller range of more streamlined European-style products and away from the gas-guzzling, boxy American cars. After the design change, Ford's US market share, which had dropped to 17 per cent in the early 1980s, climbed steadily to over 22 per cent by the end of the decade.

As a further example, Philips found that their engineering-led attention to innovation and to detail meant that they were far slower in getting a product to market than their Japanese competitors. They were following the classic manufacturing route of producing the perfect product, and then waiting for people to buy it. A change in approach meant an improvement in their overall business performance, and gave them the foundation for future development. (Lorenz 1990 gives detailed case studies of these and other design successes.)

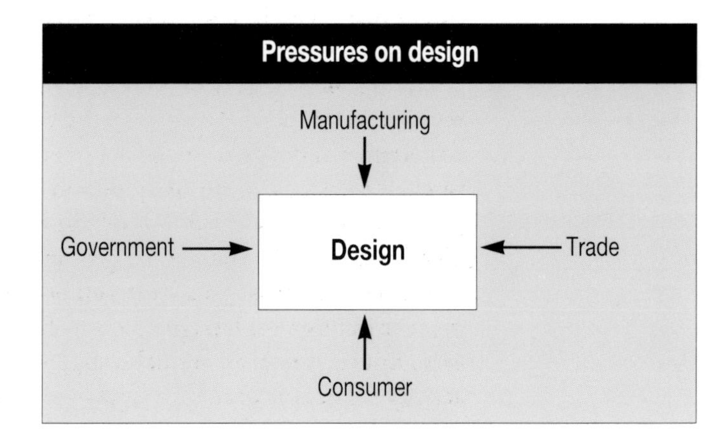

Increasingly, companies are realizing that, like Ford and Philips, survival means a conversion from a product-led company ('we make products in the way that we want to, or feel we can do, and then sell them')

to a market-led company ('we find out what consumers want or feel that they need, and then design products which will satisfy those wants or needs'). The same strategy applies to service companies, where the service is the product.

A market-led company requires a huge, constant commitment to design; to the product itself, to the environment in which it is found, and to the way in which the benefits are communicated to the target market. There have been some very encouraging results from companies who have recognized the power of design. An example is Ross Radios, who tackled the Japanese challenge head-on, and won, selling successfully not only in UK outlets, but also internationally, including Japan, because of their innovative product design.

Design management practices in UK companies

The quality of design management in companies, and subsequently the quality of the design impact, can vary widely. Good design is the result of a carefully planned and controlled programme, carried out either by in-house designers or by external consultants. All too frequently, however, the design impact in UK companies is still simply the result of random, organic development over time; inevitably, both time and budgets are dissipated.

Different types of design projects can be commissioned by several departments in any one organization, whether the work is to be done by in-house designers or external consultants. Communication design is usually commissioned by marketing departments, who sometimes fail to recognize the need for synergy between the imagery used in design with that used in other forms of communication such as advertising. All too often, press advertising looks as though it comes from a different company from that supplying the product instructions; a television commercial often fails to support the pack design, or vice-versa. Environmental design is bought by a wide range of people; the managing director, the architect, the personnel manager, and the receptionist can all be in charge of commissioning work from different suppliers. Product design is often briefed by engineers or technologists, without sufficient discussion with, or commitment from, the marketing department.

Some companies, in order to fulfil their declared commitment to design, have established design management systems with qualified design managers as senior and influential members of the company. These companies are beginning to reap the benefits of those systems through a firm control of the overall design strategy and design expenditure of the organization. Examples of such companies include WH Smith, British Telecom, Boots, Sainsbury, Courtaulds, British Airways and BAA. Frequently, companies which are able to retain a tight control on their design impact are owner-managed, such as The Body Shop and Virgin. In addition, companies which do not have their own permanently employed design managers are beginning to realize the value in retaining the services of independent design managers.

In each case, the role of the design manager is the same: to ensure that design is allowed to reach its maximum potential in the organization by meeting clearly defined and universally agreed objectives, in an efficient, effective manner. Of course, design does not work only on a rational basis; it harnesses a complex mix of elements including aesthetics, imagination and innovation.

Sadly, many UK companies still do not manage design with the same attention to detail at board level as would be given to managing other resources, such as finance or personnel. In part, this stems from a historical lack of knowledge of the potential of design, and from the absence of design as a subject in most business degree and other courses. To take two examples from the communication design discipline alone: the Communications, Advertising and Marketing Foundation and the Chartered Institute of Marketing courses currently cover every aspect of creative services as separate subjects, except design management.

However, the introduction into business schools of elective design management courses, such as those at the London Business School and Middlesex Polytechnic, will gradually change the way in which client companies address design. The Design Council runs regular seminars for design buyers, and a series of design management guides produced by the Design Business Association (published by the Department of Trade and Industry) will further the cause of effective design in British companies.

If effective design management becomes a recommended subject on every busines studies degree course in the country, and in every client company's in-house training scheme, design will be valued at the appropriate level, by the board.

How can design companies help themselves?

The design market is gradually maturing, helped by the educational initiatives which are being taken by colleges, where designers can study design management and business practices alongside their chosen design subjects. In addition, the Design Business Association and the Chartered Society of Designers are both working to establish continuous professional development training for people working in design.

Several leading UK design companies are setting an excellent example by promoting the potential of consultancy in design. Some companies such as Coley Porter Bell, Lloyd Northover, Newell and Sorrell, Sampson Tyrrell, and Wolff Olins, have employed highly qualified and experienced business managers from outside the design business to work alongside creative staff and develop the role of consultancy.

What can design companies of any size do to encourage their clients and potential clients to think of design as an effective business tool, and to give it significant financial muscle? How can they persuade the same clients of the power of design as a long-term consultancy process rather than as an ad hoc, tactical answer to an immediate problem?

One way is to read, and to encourage all staff in the design consultancy to read, as many reported case studies as possible, to learn the lessons in them, and to be able to quote them to design buyers. Some excellent case studies are available (see *Further reading*). This knowledge can then be linked to a thorough familiarity with general business developments, obtained by regular reading of the financial press.

The published results of the Design Business Association's Design Effectiveness Awards provide some pertinent examples of the design successes of a wide range of companies, including projects from Courtaulds, Boots, United Distillers, Rabone Chesterman, The Body Shop, Redland, Highland Spring, ICI, Next Retail, Grand Metropolitan, and Merck Sharp and Dohme. Submissions are judged by design buyers not only on creative merit, but on proven results in the marketplace. As a result, design companies are increasingly able to quote the case histories of the winners and the finalists in support of the case for expenditure on design.

In theory, design companies which are part of a larger communications group should be able to develop consultancy skills by benefitting from the cross-fertilization of ideas with other creative services companies in the group, such as PR, advertising or direct marketing. In practice, however, many such companies continue to operate more or less unilaterally, only taking advantage of improved financial support and useful introductions to potential new clients by other members of the group. The failure to capitalize on these seemingly obvious opportunities is part of the difficulty of implementing large-scale management change.

The best way of establishing the power of design is by working with existing clients. Carrying out a critical review of all completed design projects as a matter of course, not only when a case study is needed for an awards competition, will provide a firm base of knowledge of the way in which design has:

- met the original objective outlined in the brief
- affected the client company, product, service or brand in the longer term.

This knowledge can then be shared with all staff, however junior, and of course with all clients, and potential clients.

This approach means asking more of clients, in order to get the base data needed before the design is active, so that the effect can be measured. It is important wherever possible to isolate the effect of design from other variables, such as price, distribution, promotional activity and expenditure, and competitive activity. This, of course, requires obtaining and analysing data on those other variables, employing if necessary an external statistical consultant, and asking clients to part with post-design sales data. They currently do this for other consultants such as advertising agencies. It will mean learning much more about trade, market and consumer research, and being brave enough to learn that, sometimes, the new design will have no effect, or even a negative effect. Assembling and analysing this information may even require some expenditure on the part of the design consultancy.

Unlike other business disciplines (such as advertising, direct marketing, in-house training or cash flow management), which have been statistically proven over a long period of time to be effective, there is still a great deal of pioneering work to be done in the area of design effectiveness by the whole industry. Leading advertising agencies, for example, estimate media effectiveness on the basis of past sales results; they build econometric models to judge the effect of different levels of activity or expenditure; they can estimate the number of OTS (opportunities to see) which a given representative of the target market might have for any particular advertisement, and thus judge the frequency of viewing needed; they measure the effect of creative solutions both qualitatively and quantitatively through market research; they test campaigns regionally in order to measure effectiveness. In contrast, there are very few design companies currently involved in this sort of activity for their clients.

Despite design companies' surprising reluctance to ask, most clients are only too willing to assess the effect of expenditure on design. It is, after all, in their interest. The longer the relationship with the client, the easier the evaluation process becomes. Proving that design works takes time, effort, and commitment, but it pays dividends, which may in the long term include securing the future of design consultancy.

Conclusion

Design does not start on the drawing board; it starts in the brain, and often in more than one brain. Good design will almost always lead back to a good client. Specifically, it will always lead back to a good and fertile brief. A good client is one who knows what he or she is trying to achieve, not in terms of execution, but in terms of objectives. A good client also knows the capabilities and the limitations of design.

Design is improved, not hindered, by intelligent discussion and development and market testing. Sony, which is an exemplary company committed to design, claims that it can learn more about a product in two months in the marketplace than in two years in the laboratory. As a result, it anticipates consumer wants or needs, creates a solution in the form of a product, often with a development period of only six months from idea to launch, and then keeps on improving that solution as a result of experience in the marketplace until the product is discontinued.

Styling, as opposed to design, is often criticized, and it certainly should not be used to cover metaphorical cracks, mould or rot in the product or service offer. However, styling or re-styling can be valid; it can successfully differentiate the product or service in question from the competition, albeit on a short-term basis.

Real design, however, is more than skin-deep; it works from the inside towards the outside. It achieves an effective and efficient result, long term. It improves business performance, and, ultimately, business profitability.

Further reading

Bernstein, David (1986) *Company Image and Reality*. London: Cassell.

Burall, Paul (1991) *Green Design*. London: Design Council.

Design Business Association (1989, 1990) *DBA Design Effectiveness Awards publications*. London: Haymarket Publishing.

Design Business Association (1992) *DBA Managers Guides*. London: Department of Trade and Industry.

Design Research Company (1991) *Design Buyers Report*. London: DRC.

Fairhead, James (1988) *Design for a Corporate Culture: A Report*. London: National Economic Development Council.

Gorb, Peter (ed.) (1990) *Design Management*. London: ADT.

Gorb, Peter and Schneider, Eric (eds) (1988) *Design Talks!* London: Design Council.

Lorenz, Christopher (1990) *The Design Dimension*. Oxford: Basil Blackwell.

Mackenzie, Dorothy (1991) *Green Design for the Environment*. London: John Calmann & King.

Olins, Wally (1989) *Corporate Identity*. London: Thames & Hudson.

Olins, Wally (1990) *The Wolff Olins Guide to Corporate Identity*. London: Design Council.

Pilditch, James (1987) *Winning Ways: How Companies Create the Products We All Want to Buy*. London: Mercury.

'A verbal agreement isn't worth the paper it's written on.'

attributed to Sam Goldwyn

Michael Munday/Sharp Practice

12 AGREEING TERMS OF BUSINESS
David Jebb

Terms of business are intended to be a framework within which designers, clients and suppliers work. Terms of business need to be established between all of the parties concerned before any work begins, so that each party to the proposed transaction knows in advance the rules by which the game into which they are about to enter is to be played. This is important because the rules by which any one company always plays frequently clash with the rules by which the other parties to the business always play. It is as well to get contradictions sorted out before the game starts.

Trying to establish retrospectively the terms of business under which some completed transaction was done is normally the prelude to a dispute; general terms of business are intended as the first step towards avoiding this.

Project-specific terms of business

Before a design consultancy starts to work for a client, or before a freelancer or a supplier starts to work with a design consultancy, it is also necessary to agree project-specific terms of business. Whereas the general terms of business specify the general rules of the game, these specify the nitty gritty of what that game is to be:

- what is it that each participant is providing, to whom, and when?
- what is it that the parties involved are expecting to receive?
- within the framework of the transaction who will be responsible for what?

It is possible to work by word of mouth, and on the backs of envelopes, but this will not provide an orderly framework within which to conduct business. A frame-work allows time to be used productively; word of mouth and backs of envelopes guarantee that a lot of time will be involved in disputes and litigation, to the benefit of the legal profession, but to the detriment of the design consultancy and its clients.

Drawing up general terms of business

Never try to draw up general terms and conditions of business without professional legal help. Use a solicitor who is experienced in commercial law, and, if possible, is already familiar with the design business. Working together, design consultancy and solicitor should seek to draw up general terms of business which cover:

- legal identity of the consultancy
- title (ie legal ownership) of the work
- the limits of the design consultancy's responsibilities
- the use of the work
- design credits

- payment terms
- interest on late payment
- interpretation (see below)
- copyright, moral and intellectual property rights
- governing law.

How all of these things are expressed, and exactly what should be included for any particular business, are matters arising directly from the company's circumstances and objectives, and must be dealt with accordingly. There are no 'tablets of stone' on terms of business in design consultancy, but it is in everyone's interests to operate within clear, fair and workable guidelines:

- the terms of business must make it clear who is offering to do the the work; the potential client needs to know the legal status of the consultancy
- legal ownership should not be confused with physical possession; the legal owner holds the title (eg the title deeds to a house)
- establish the point at which title and property rights to goods and services supplied will be transferred (this applies to supplier-to-consultancy transactions and consultancy-to-client transactions)
- the consultancy may want to reserve the right to recover all costs incurred in recovering overdue payment from the client
- 'interpretation' involves clarifying the legal meaning of words and phrases used in the terms of business
- the terms of business should state which country's legislation will cover agreements made (particularly important in the light of increased overseas work).

In a number of instances the terms and conditions which the design consultancy wishes to adopt for itself (eg in relationships with suppliers) will be different from those it wishes to apply in relationships with clients.

Establishing a project-specific agreement with the client

Outline brief

An outline brief should, ideally, be prepared by the client before discussions on the project-specific agreement begin, and this is likely to form the basis for those discussions. Subjects that ought to be covered include:

- background information on the client organization
- the subject, context, and nature of the design project
- detail on the marketing and business background
- clearly stated project objectives
- description of the target audience
- timing
- the names and roles of the people in the client organization who will have the authority to approve the work at each stage
- budget
- constraints
- environmental considerations.

Incompatibility

Incompatibility can arise between the terms and conditions of the consultancy and the client. The negotiation may start from a basis where the consultancy's general terms of business and the client's terms of business are not compatible. The client is the buyer, and the consultancy the seller, and each party will have drawn up general terms of business which protect and support their own interests.

The project agreement negotiation has to move from the general to the specific, in order to agree what terms of business will relate to the particular project.

Problems arise if consultancy and client send each other documents which, in effect, say 'the above general terms of business are set in concrete, and will govern the conduct of anything we do together unless we agree to any change, in writing, before any work commences'. If the initial documents are incompatible, and they are left as such, this can lead to what lawyers sometimes call 'The Battle of the Forms' where both sides are endeavouring to establish whose terms and conditions take precedence in the event of a dispute.

What makes a contract

For a contract to exist there needs to be:

- an offer
- an acceptance
- a consideration (ie an agreement to transfer something of value such as money, goods or services).

If the client's general terms of business include items unacceptable to the design consultancy, or vice versa, alterations that make them acceptable to both parties must be thrashed out before a contract is made. If this is not done, the resulting contract may be invalid as it is quite possible that a court would decide that all that has happened is that the consultancy has made an offer, and the client has made a counter offer. There has been no acceptance, and therefore no contract.

The parties must agree the consideration as part of the negotiation. Once the offer is accepted, a contract is made. The majority of contracts made in the design sector are verbal in the first instance, the three conditions that go to make a contract having been fulfilled at a meeting or over the telephone. This does not make what is now the post-contractual exchange of documents any less important.

Precedence between general terms and conditions

Once a contract exists, if there is a dispute over whose general terms of business apply, the general view is that the terms which take precedence are the last ones despatched before the conclusion of the contract. This is, a legal 'grey area' which is best avoided.

Elements of a project-specific brief

The aim of the project-specific agreement is to reach a well-defined and recorded arrangement with the client so that:

- there is a mutual understanding of what both parties are setting out to achieve; this is normally contained in a written brief
- the agreement which has been reached on price is set out clearly; the consultancy informs the client of what charges are to be passed on as rechargeables and/or disbursements, and supplies either costs or estimates for this
- there is an agreed timetable both for work and for payment
- both parties know the position on title to the goods and services to be provided
- consultancy and client are in agreement over any legal rights related to the project
- both parties are agreed on how changes to the client brief are to be handled
- an agreement is in place for procedure should work be rejected
- mutually acceptable procedures have been agreed concerning confidentiality
- agreement has been reached on the availability of file copies to the consultancy.

Client brief

In order to arrive at this point in their discussions, consultancy and client will almost certainly now have refined the document setting out the aims, objectives, extent and nature of the work to be carried out, together with any other relevant supporting information. This is now the client brief.

It does not matter whether the brief is drawn up by the client or by the consultancy, but it is very important that it is done, and that both sides are agreed on the content. The brief will form part of the design consultancy's proposal document, and the proposal is the most usual method of drawing together all the elements of the project-specific agreement.

Once the final project agreement has been reached and committed to writing, make certain that it can be altered only with the mutual consent of authorized people representing the parties involved; their names should be included in the agreement.

Design stages

The established stages of a design project are normally regarded as being:

- research, strategy or feasibility
- design concepts
- design development
- implementation

with the first two stages often being rolled together into one.

There is, however, an increasing trend for clients to commission a separate pre-concept stage to establish the feasibility, cost, or direction of a design project. This should be regarded by the design consultancy in the same way as is any other stage, with fees and a programme of work being agreed before the work is started.

The next section of this chapter deals with the various types of work which the design consultancy will have to carry out before it is able to present the detailed proposal document; this will include:

- considering how the work will be handled, and who will work on it
- refining all elements of the project specific agreement
- project costing and preparing a detailed quotation.

Confusion can arise between the proposal which is presented at the pitch stage, which will include only general indications of how the consultancy would tackle the work if appointed, and the detailed proposal which is required to regulate things once a consultancy has been appointed.

Changes to the client brief

Probably the most disputatious part of agreements between design consultancies and their clients concerns changes to the client brief. To minimize the possibility of dispute, the consultancy should insist on all changes being requested in writing. If the proposed changes will cause alteration to other things which have previously been agreed (eg the delivery date, or the price), the consultancy should inform the client of this in writing, for example explaining what the new delivery date and price will be, and insist on acceptance of the new arrangements before proceeding. Even if there is no knock-on effect from the requested change, the consultancy should confirm (in writing) receipt, and acceptance, of the request to avoid any further confusion.

Changes to the client brief may be requested at a point where the project is time critical. No matter how urgent the changes, or how rapidly they must be implemented, never omit the written agreement stage; fax machines may help here.

Termination of contract

The design consultancy or its client must have the ability to bring the contract to an end if the other is in serious default or becomes insolvent. Serious default may include non-payment of fees by the client, or failure by the client to give proper instructions, or failure by the consultancy to execute the work to a reasonable standard within a reasonable time. Standard terms will normally specify that written notice of termination is required. Termination by this method is possible even where there is no default, for example for commercial reasons, in which case a reasonable period of notice (two or three months) must be given. This type of termination requires the paying party to pay anything which is outstanding up to and including the period of notice, as well as outside costs incurred on the project. Provision should be made for what happens to work done up to the point of termination.

Project costs and project prices

Before a consultancy is in a position to finalize any project agreement with a client, it is necessary for the project to be costed and priced. Once broken down into its component parts this daunting prospect becomes digestible. Developing the costs and calculating the price includes the following:

- how much work will be done in-house using own staff
- staff hours by 'designer weight' or by function for each part of the work
- how the project will be managed
- how any necessary subcontractors and suppliers will be selected
- what basis for payment is to be proposed to the client
- whether print costs, and any similar items, will be bought in and sold on to the client, or whether the consultancy will act as agent for the client
- the method of charging for disbursements.

Using an approved supplier list

Buying in work and services can be a weakness in the design sector; a relatively small expenditure of management time and effort in this area can be cost-effective.

It is normally good practice for a design consultancy to keep an approved list of suppliers for goods and services which are required regularly. The development of a close working relationship between the consultancy and its approved suppliers will normally lead to a higher level of service than ad hoc ordering, as well as giving the consultancy tighter control over both quality and cost.

To be fully effective, an approved supplier list must be kept alive. Companies which wish to be considered for inclusion should be allowed to pitch for the business, and companies which under-perform should be removed. If this is not done the competitive edge achieved through the use of an approved list is lost, as the longer-term supplier companies may become complacent and lose their edge on pricing.

Agreeing prices with suppliers

The two methods most commonly used for agreeing prices with suppliers are specification and tender, and negotiation.

The method selected depends on what is being bought, its value, and the number of competing sources from which the product or service can be obtained. If few options are available, the consultancy will probably have to negotiate. With a wider choice, (eg with a large and technically unsophisticated production run that is not time critical,) specification and tender is likely to be the most cost-effective route.

Such a tender is likely to be by invitation, and, in many cases, the consultancy will be able to agree the keenest price of all by following the route of specify-tender-negotiate, but, if this is the intention, it should be made clear to the competing suppliers when they are invited to tender.

During negotiation, the consultancy should also give attention to agreeing the stage at which title and any other rights (such as copyright) will be transferred. A hard bargainer may ask for these to be transferred on delivery, although suppliers can insist on retaining title and property rights until payment has been received. The consultancy must ensure that it receives title and property rights before passing either or both of these on to a client.

Where a consultancy is able to make the choice, it must offset the advantages of holding a creditor's money for an agreed period of time against the level of discount which may be offered for prompt payment. A cash rich company should be able to negotiate discount terms for seven- or fourteen-day payment, which is substantially more beneficial than all but the most extended creditor financing.

Preparing the quotation for the client

Remember that an estimate is the non-binding and approximate judgement of an amount, whereas a quotation is a statement of the price for which, within agreed conditions, work will be undertaken. In developing a quotation for a client there are three major cost component groups:

- fees
- rechargeables
- disbursements.

The fee element of a project will normally be calculated on the basis of a fixed fee, hourly or daily rates, or on the basis of a retainer.

A quotation for a clearly defined and specific assignment will normally be calculated on the basis of a fixed fee.

Fixed fees

Fixed fees are now the rule rather than the exception. Commercially aware buyers of design want to know before they commit themselves exactly how much each stage of each project will cost; they want all risk of uncertainty to be borne by the design consultancy. The calculation of a fixed fee is theoretically straightforward, but the final quotation to the client will need to be tempered by a number of commercial considerations, such as workload, the consultancy's keenness to undertake the project and/or to work for that particular client, and how prestigious the involvement will be.

Commercial reasons for altering a fee calculation must be a matter of judgement by the individual consultancy, but should always be given careful consideration. Do not confuse pricing and costing, and always calculate:

- what the work will cost you to do
- what the fixed fee for that work should be.

If offering a lower quote the consultancy must establish the difference between the new price and the full price, and relate the quotation to the cost of the time that will be taken to generate those fees. (See chapter 9.)

Hourly and daily rates

In the past, clients accepted hourly or daily rate fees because the time element of the designer's work was regarded as being difficult, or impossible to calculate. Nowadays most clients wish to know, in advance, the total cost for which they are making themselves responsible, and are becoming increasingly insistent that all work is done on a fixed fee basis. This means that the risk attached to forecasting how long the work will take is borne totally by the designer, and this must be taken into the account when the fee quotation is being prepared.

An hourly rate may still be an acceptable fee basis to a client for experimental work, or for work carried out within an established designer/client relationship.

Charging for extra work

Hourly rates may also be relevant where additional design work results from changes to the client brief. Any additional work should certainly be charged for, either on an additional fixed fee basis, if it is possible to calculate the additional time needed for the implementation of the changes, or at an agreed hourly rate. Clients tend to underestimate the amount of additional work that results from their making changes to the brief, and often suggest the cost should be met from the profit margin as their budget is already fully committed. If changes are agreed and confirmed in writing, as discussed earlier, much misunderstanding in this area can be avoided.

Retainers

A retainer fee will normally only be acceptable to clients if they want regular input on design matters, or if the consultancy is undertaking work of a repetitive nature (eg a company producing a quarterly house magazine for a large organization). In this situation try to negotiate a retainer fee covering the repetitive part of the design input, and agree an additional fixed fee for each issue to cover the non-repetitive work.

Royalties

Clients sometimes ask for work to be carried out on a royalty basis, particularly in product design. This can be a lucrative arrangement, but requires caution, as the prospective client can be over-optimistic on sales estimates. If tempted, consider a split deal, with lower fixed fees, and a royalty formula for the balance.

Exclusivity

The client may wish to achieve exclusivity on some aspect of a consultancy's practice. A consultancy entering any form of exclusivity arrangement must charge a rate which is not only commercially viable for the work involved, but also compensates for other work which must be turned away. The agreement must also be reviewed at regular, and pre-determined, intervals.

Current charging practice

On the premise that the most valuable items the design sector has to sell are original conceptual thought and design solutions, the current charging structure is largely the wrong way round. Unless the job is top-loaded (ie the majority of the fee charged in the early stages) conceptual thought and the design solutions will be provided at a subsidized price, while the more mechanical work of development and production are charged out at a much more economic rate.

Rejection fees

Designers must always negotiate a formula for rejection fees before starting to work for a client, so that if the project is terminated a satisfactory level of payment will be made to cover the conceptual and early stage work. If work is rejected, both sides will need to refer to the brief and the project agreement, in order to establish a fair level of compensation for the work done, taking into account whether the failure derived from execution or taste.

Pre-payment of fees

From each and every client the design consultancy should ask for a substantial percentage of the fee income for each stage to be paid when that stage is authorized. A number of consultancies from a variety of disciplines now ask for 50 per cent, but this figure should be regarded as being negotiable. The fees quoted in any offer should cover the time which will be spent by the consultancy on design. Items such as illustrations, photography and production, should be included as a separate charge.

Sale of rechargeables

It is important that the prospective client understands in advance that these rechargeables will be billed separately from, and in addition to, the fees quoted. The client will probably want a fixed quotation for rechargeables before giving the go-ahead for the project, and the designer should include a prudent contingency in this price. If it is not possible to calculate and quote a price, agree with the client what items will be invoiced as rechargeables, and try to provide likely maximum and minimum figures as guidelines.

Disbursements

Expenses which are incurred specifically on a client's behalf, such as project travel and subsistence should be treated in the same way as rechargeables, and billed to the client with whatever mark-up the consultancy applies. Items to be classed as disbursements should be agreed with the client before the start of the project, and the quotation should include an estimate for these.

To avoid unnecessary calculations a number of consultancies quote this element as a percentage of total fee income for the work. The percentage to be applied is calculated from a statistically valid sample of recent invoices. A charge for disbursements at 5 per cent of fee income is not uncommon.

Acting as the agent of the client

Some consultancies prefer the client to be legally responsible for bought-in services, such as production, with their only involvement being as agents. This type of arrangement can enable a company to handle production runs of a size it could not manage without the backing of a substantial client, and minimizes the consultancy's exposure to default.

There are arguments against acting as agent for the client:

- the consultancy will only be paid a handling charge which is likely to be substantially less than the income from the sale of equivalent work as rechargeables
- although the contract will be between the supplier and the client both of these parties will still use the consultancy as the intermediary, which can be responsibility without control or reward.

The best commercial practice is normally for design consultancies to buy the production and then sell it on to the client, although there are likely to be exceptional instances, for virtually all design consultancies, when it makes more sense to act as the agent for the client.

Agreeing terms of business is one thing. Getting paid for what has been done may well be another.

Credit checks on prospective clients

A consultancy considering undertaking work for a new client should take basic precautions by asking for trade and bank references. It is also worthwhile asking a credit agency for a credit report on the client company, and if possible, talking to other companies who have worked with that client. Many of the country's largest companies are amongst the slowest payers, and these informal enquiries can reveal this information more effectively than trade, bank, or credit agency references.

Market conditions and the 'custom and practice' associated with each sector of the design industry will determine what percentage of the fee income may be apportioned to each stage of the project.

Pro-forma invoices

If dealing with a new and unknown client whose references and trading record seem to be at all 'iffy', raise a pro-forma invoice for a substantial first-stage payment and insist on this being paid before starting work on the project.

Putting the proposal to the client

At this stage, the consultancy is now in a position to pull together all of the component costs which will go to make up a quotation for a stage or for the whole project, and to put the detailed proposal into a written form for presentation to the client. A proposal letter or document must cover all that needs to be said in an accurate and precise manner, without becoming unduly wordy.

As the proposal is the offer, and therefore one of the three events that have to take place before a contract comes into existence, it is essential to get it right. If there is a dispute between the consultancy and the client at a later date, this document will be of critical importance.

Calculating the price

Before sitting down to write the proposal, it will be necessary to pull together and collate all of the following 'raw' calculations which contribute to calculating the fee and cost quotation:

- in-house design time at standard charge-out rates
- selling price of labour rechargeables, including any freelances required (fees + company's standard mark-up)
- selling price of material rechargeables (purchase price + company's mark-up)
- selling price of production (cost based on written quotations from supplier + company's mark-up)
- selling price of disbursements (actual or estimated cost + company's mark-up).

Once this stage has been achieved, it is necessary to consider the actual price to be quoted to the client, adjusted in the light of its actual value in current commercial terms. Perceived value is an important commercial consideration; there may well be instances where the proposed calculated costs, plus margins, produce a price which is below market value for the real service which is being offered. In such cases, consideration should be given to adjusting the price, perhaps by increasing the rates charged to design time to reflect the contribution of creative and innovative thought on the eventual commercial impact of the project.

Writing the proposal

A sample structure for a proposal document could be as follows:

1 Title page

Give basic reference information about the document: the name of the project; the name of the client; the name of the design consultancy submitting the proposal; the name of the person who prepared the document; the date it was written.

2 Introduction

This is a short summary of the aims, objectives and background to the project – a short form digest of the brief. It can set the tone for the proposal, by confirming the consultancy's understanding of and attitude towards the project.

3 The brief

If possible, it is a good idea to include a complete version of the agreed brief in a proposal document. The proposal is related to the brief, so it helps to have it there for reference and clarity. However, sometimes the brief will not have been fully developed at the time of preparing a proposal; if this is the case, say so, and explain how the brief would be worked up and agreed in a future working relationship.

4 The approach and method of working

This section explains to the client how the work would be tackled. It may include types of work to be undertaken, and will explain the nature and extent of the work which the consultancy is offering to carry out, and it forms an important part of the legal offer, becoming part of the contractual agreement if the work goes ahead. Do not assume that the client understands the different stages of the design process, and the terminology involved; spell it out very clearly, and then make sure that the recipient understands what is being said. If possible this section should also include an outline timescale on the project; it may not be possible to quote actual dates, but give an indication in terms of weeks from the start date. Also include how long the offer will remain open at the fees stated.

5 The team

It is a good idea to set out both the types and levels of people to be involved in the project – so that the client can see who will be doing what – and also to include details on the senior members of the team and how their previous experience relates to the work required. This also constitutes part of the offer, and people must not be included if they will not be available to work on the project in the role described.

6 Design fees

As explained earlier, this may be a fixed fee, hourly rates within a budget, or a percentage of a contract. The proposal should set out the fees to be charged for each stage of the project, and explain on what basis they are charged, together with details of what is included/excluded. It is also sensible to include the hourly or daily rate to be applied for any additional work required for agreed changes to the client brief.

7 Rechargeables/disbursements/production costs

The proposal should include details of items to be charged separately under each category, together with either quotations or maximum/minimum cost guidelines.

8 Reference to general terms of business

In order to arrive at a watertight legal offer, the proposal must either include or make direct reference to the consultancy's general terms of business. Check that in this way necessary additional elements such as rejection fees and arranging for invoicing and payment are included in the proposal.

9 Confirmation of acceptance

It is necessary to ask for written confirmation of the client's acceptance of the proposal (offer, acceptance, consideration); this is often done by including a second copy of the document to be signed and returned to signify agreement.

The purpose behind reaching agreement on terms of business is to avoid confusion, and to make certain that consultancy and client, or consultancy and supplier, are singing from the same hymn sheet. Don't leave things to chance. Ensure that all of the items on the checklist (see above) have been covered before putting the arrangements into writing and starting to work with the client. This route clears the way for spending time together productively and profitably, instead of trying to put out an endless succession of bushfires.

Checklist for finalizing an agreement with a client

- ☐ Have I agreed with my client precisely what I am going to do for him or her?
- ☐ Is there any aspect of the brief on which I require further clarification before I start work?
- ☐ Have we agreed a date for the completion of this work?
- ☐ When will the title to the goods and services I am going to provide be transferred?
- ☐ Have we reached agreement on copyright?
- ☐ What about other intellectual property rights?
- ☐ Does my client know and understand the limit of my responsibilities?
- ☐ Am I going to be given the design credit for this work?
- ☐ What limits have we agreed on its use?
- ☐ Has my client agreed the price that he or she will pay me for this stage or project?
- ☐ Have we agreed the terms of payment?
- ☐ Did we agree rejection fees?
- ☐ Does the client know that he or she will be charged interest on late payments?
- ☐ Has he or she agreed that I will have the right to recover any costs which I might incur in recovering overdue debts?
- ☐ Are the terms and conditions under which I am undertaking this work clearly understood by the client as well as by me?
- ☐ Does the client understand that he or she will be charged for any extra work that results from making changes to the client brief?
- ☐ Have we agreed how such charges will be calculated?
- ☐ Does a legally binding contract exist between us?
- ☐ Has everything been confirmed in writing?
- ☐ Are there any loose ends which I would like to tidy away before I start?
- ☐ In the event of a dispute arising between us, under what country's laws is this agreement to be interpreted?

Further reading

Kennedy, Gavin (1991) *Everything is Negotiable.* London: Arrow.

Lloyd, Stephen (1991) *The Barclays Guide to Law for the Small Business.* Oxford: Basil Blackwell.

Parris, John (1988) *Making Commercial Contracts.* London: BSP Professional Books.

Stones Porter Solicitors (1990) *Designers and the Law.* London: The Design Business Association & Stones Porter Solicitors.

Jonathan Field/Sharp Practice

'It is a fight; in order to win you must do better than your rival in every minute point, in the run of the whole thing and in all the details.'

Le Corbusier

Towards a New Architecture

13 PROJECT MANAGEMENT

David Rivett

Every design business wrestles with the essentially intermittent nature of its relationships with clients. Although there are clients who pay retainers, they are a minority; for the most part the designer cannot rely on a continuous stream of work or a guaranteed level of fee; it comes in packets called projects. It should follow that these packets are cherished and nurtured, and yet poor project management is probably the biggest single cause of client defections. The importance of good project management in ensuring that clients remain happy and loyal cannot be over-emphasized: 'You are only as good as your last project.' There are five keys to successful project management:

- clearly defined responsibilities and leadership
- a good brief
- thorough planning
- clear and frequent communication
- proper controls and monitoring.

Project responsibility

The first issue to be decided is who is responsible for the project? From the moment a proposal is written, both client and consultant must know exactly who is in overall charge and who handles the day-to-day running of the project.

The brief

In order to submit a fee proposal the designer must have received some form of briefing from the client (see chapter 12). However, perhaps because of clients' understandable reluctance to circulate confidential material to a number of competing consultancies, or perhaps because the brief can be interpreted in a number of ways, it may need some clarification in order to avoid confusion later in the project.

The brief review

The brief review should take place at the beginning of the project and is an opportunity for the design team and the client to make sure that all the fundamental assumptions and objectives are understood and agreed. The design team can question the brief and probe the thinking behind it.

Because designers may well be questioning assumptions which are grounded in an intimate knowledge of running a business, this is a time for great sensitivity. Designers should seek to understand the client's organization, problems, strategy, operational issues, financial objectives and budgetary constraints. Only when this understanding has been built up will the design team be properly equipped to lead the client forward or possibly even to counsel caution. Some particularly valuable creative work has resulted from designers helping clients to rework the brief.

Research

Either because planned from the outset or as a result of the discussion and questioning during the brief review phase, it may be that some pre-project research and client familiarization is required. Chapter 17 gives guidance on research principles, but from a project management standpoint it is important to know:

- who is doing the research
- what form of output will be most useful
- how long it will take.

Since research is a feedback mechanism, it is important to allow time to undertake the fieldwork, collate the findings and feed the conclusions back into the brief.

Finalizing the brief

When the project is straightforward or the original brief comprehensive, the brief review may be a short meeting which can be summarized in a memo confirming the appropriateness of the initial proposal. If discussion and research have generated new ideas and widened the scope of work, then a new brief may need to be written.

Whichever is the case, the brief agreed at this point provides the benchmark against which the design team's work will be assessed. It is therefore vital that client and designer are in complete agreement about the objectives, constraints and scope of work when the brief is finalized.

It is important that clients realize that any changes are likely to result in at least some abortive work and quite possibly additional fees. Thus the need for clarity and agreement at this stage cannot be overstated.

Cementing good relationships

Every good proposal will set out the stages through which a project will pass. Taken together these constitute the design process. The chart (opposite) provides a list of possible project stages, although not all projects need involve all stages.

Even amongst design consultants there are variations in the terminology used to describe a particular phase or stage, and it is not surprising that many clients are not sure what to expect and when to expect it as the project progresses. It is important to explain terms such as concept design, sketch models, detailed design, and even artwork, to tell the client what will be delivered, who will participate and how.

By getting the client comfortable with the consultant's terminology and process, misunderstandings and disappointments will be minimized. Encourage the client to visit the studio so that he or she has a feel for the physical process and gets to know the team. The client will feel greater ownership of the project and be better able to explain what is going on to colleagues.

Establishing the right channels of communication

It is important to identify project champions with a special commitment to the work both within the client company and the consultancy at the earliest possible opportunity. On the client side a single point of contact should be identified to organize all internal communication and liaison. If this is not feasible, each contact must have a clear non-overlapping focus; it is embarassing to have the managing director and marketing director both trying to set overall direction and contradicting each other. The creation of a small project steering group could be encouraged if this can defuse internal politics and involve potential project adversaries.

In a small design firm where a partner both makes the pitch and runs the project the champion is obvious, but as both projects and firms become larger and more people become involved there is increasing scope for confusion.

The client must be given clear channels of communication for the arrangement of meetings, for routine progress chasing, for the approval of changes to the brief and for overall project strategy and quality. If client handlers are involved, they should know exactly where their remit extends and when they should involve principals.

Frequent communication must be encouraged, by staying accessible, trying not to change the team, and providing regular progress reports. All project meetings should be minuted and minutes promptly circulated. Invitations to other specialists and professionals in the team should not be forgotten.

Whenever possible, and particularly when commencing a complex long-term project, it is useful to evaluate the relative standing of contacts within the client organization. The unconverted, the threatened and the powerful should be identified and efforts made to involve them. An extensive research or familiarization phase can be a good opportunity to talk to people who will have little day-to-day involvement in the project, but can be instrumental to its success or failure.

The design process - a simplified summary of the main stages

Graphics

- review the brief
- client familiarization
- research
- finalize the brief
- agree responsibilities
- concept design
- concept presentation/review
- agree concept(s) for development
- design development/ detailed design
- mock-ups/dummies
- consumer research
- design refinement
- final presentation
- design freeze
- typesetting
- artwork/photographs
- proofing
- paste-up
- printing/converting
- project review

Product

- review the brief
- client familiarization
- research
- finalize the brief
- agree responsibilities
- concept design
- concept presentation/review
- agree concept(s) for development
- design development/ detailed design
- prototypes
- prototype evaluation
- production planning
- design refinement
- final presentation
- design freeze
- pre-production prototype
- tooling
- manufacture
- project review

Interiors

- review the brief
- client familiarization
- research
- finalize the brief
- agree responsibilities
- concept design
- concept presentation/review
- agree concept(s) for development
- design development/ detailed design
- interim presentation/models
- design refinement
- final presentation
- design freeze
- working drawings
- tendering
- shop drawings
- pilot fit-out
- pilot evaluation
- further refinement
- fit out
- hand-over
- project review

Liaising with other professionals

Where other professionals are to be involved in the project, either at the client's instigation or as part of the design team, they should be involved early. Ensure that they each nominate a contact person and circulate these contacts with all relevant meeting notes. Other professionals should be welcomed as contributors not tolerated as a necessary evil, and it is sometimes helpful if the designer suggests specialists with whom he or she has collaborated successfully in the past.

However tempting, designers should not try to score points off other professionals but instead talk to the client if it seems that serious overlap is developing. The client should be asked to spell out the limits of each party's brief and eliminate the overlap.

Where the client has appointed a number of specialists (perhaps an architect, interior designer, graphics team and quantity surveyor) it is essential that the client co-ordinates their contributions or acrimony will almost certainly ensue. It can be tempting to assume the role of lead consultant, but unless specifically appointed to this role the designer should be wary (and even if appointed tread carefully).

Meshing with the client

There are two aspects to the creation of a sympathetic and robust linkage with the client. The first has already been mentioned under channels of communication, and involves sensitivity to the client's internal structures, politics and culture. Good consultants learn to navigate these with great skill, but remain aware that it is virtually impossible to please all of the people all of the time. A fair proportion of projects will generate serious disagreement at some stage and it may be necessary to make a stand on principle with the risk of losing a job if not a client.

The second aspect concerns the need to respect an established house style or to collaborate with another consultancy. This can be a source of friction if a consultancy is asked to conform to corporate guidelines or existing practice which it feels are unsatisfactory or inappropriate.

Many clients relate with distress the arrogance of consultants who although asked to design, say, a brochure, have insisted that only a completely new corporate identity will provide the appropriate context for the new print. While understanding the drive for quality which leads many consultants to take this stand, one should be careful of appearing to sweep aside questions of cost and tradition which may weigh heavily with the client. This type of conflict can be avoided if enough time is taken at the proposal stage to ascertain the client's current standards and the degree to which these will be binding.

In working with other design practices appointed by the client, the client's judgement must be respected, and harmonious collaboration sought. Should a disagreement arise, it is for the client to decide with whom he or she will proceed, and it is not the place of either consultant to cast aspersions on the other, whatever they may feel. Win/win negotiating techniques can be very useful here (see chapter 9).

Confidentiality

Consultants are privileged to have access to much commercially sensitive information about their clients, both directly related to projects and of a more general nature. The confidentiality of this knowledge must always be respected and impressed on staff. Before disclosing any information about a client or project which is not already in the public domain, the client's written consent must be obtained. The right to withhold such consent is entirely at the discretion of the client.

Planning

Once the brief has been confirmed, the consultant's project manager should list all project tasks for which the consultancy will be responsible, and all those which will impinge on the consultancy's work.

The client and consultant may well work jointly on the master list of tasks for the whole project.

Establishing responsibility

Once the complete scope of the project is set out in detail, responsibility for each task can be agreed. Much of this will be set out in the proposal document, but, particularly in complex multidisciplinary projects, there are numerous cracks down which small, but often key, tasks can fall. If a task is not specifically assigned never assume that it will get done. It almost certainly will not.

If the consultancy is part of a large project team, it is important that the client issues a full list of tasks with assigned responsibilities and agreed timescale to each team member. If such a project is then spread around several departments of a large practice, the project leader must also issue such a schedule internally. This becomes the master plan for the entire project.

Team composition

It is likely that at least in outline the team appropriate to a particular project will have suggested itself at the proposal stage. That team needs to be brought together and fully briefed; all team members should have copies of the proposal, and understand the client background, research findings and project timetable.

If there is flexibility in the choice of team, it should be looked at from a client perspective, and chosen with a view as to who will establish the best rapport with the client as well as contributing relevant specialist knowledge and experience.

The team should, as far as can be predicted, be chosen to provide continuity throughout the project. If it becomes necessary to change people with whom the client has regular contact, these changes should be explained at the earliest opportunity and in person.

If, because of the shifting emphasis of a project over time, it is planned to change team composition, this must be explained at the outset. In particular, the continuing involvement of principals should not be promised if this is neither practicable nor necessary; failure to deliver in this area is a major cause of client dissatisfaction.

If there are changes in the client's team, it may be necessary to fire-fight. Losing the project's champion can stop a project in its tracks, and a virtual re-pitch may be required to get things moving again under a new manager who will instinctively want to question and change something he or she has inherited.

Time allocation and budgets

In principle this should be a simple process of allocating budgeted time amongst the team members to achieve the project tasks in the timescale promised and at the cost anticipated in the proposal.

One common source of difficulty is that proposals are made in pounds and teams work in hours. Ideally the proposal should be drafted in terms of hours needed to do the job, and be converted to pounds to establish the fee, but within the teams only hours should be used for project control.

The timing and team input of each phase of the project needs to be scheduled with care and plotted on a project timetable. Where a person is involved in several projects their time must not be committed twice. Even with all this planning, team allocations and workload should still be reviewed weekly and people and projects juggled to accomodate the unforeseen and to keep people busy.

When putting together the project programme realistic time should be allowed for decision-making. If possible a client's decision-making processes should be discussed or assessesed before drafting the proposal. Delay in obtaining decisions can destroy the timetable and leave people sitting around idle.

Critical path analysis

On large complex projects, and particularly in the implementation phase, there will be a web of interdependent activities. Certain of these activities will define the critical path, that is the shortest time, if everything goes to plan, in which the programme can be completed. A delay in any activity on the critical path will delay completion. When working towards a given opening date for a shop, hotel, or exhibition for example, this can be a serious problem. The designer is never working in a vacuum, and must communicate constantly and immediately regarding slippage or acceleration of the timetable so that this can be allowed for.

A programme chart should be drawn up setting out all project activities and highlighting the critical path. At its simplest this can be a bar chart plotting activities against time, but for more complex projects a series of techniques known as project network analysis has been developed. The subject deserves a chapter to itself, but the basic principles are quite simple, and there is plenty of accessible material on the topic and numerous pc computer programs which help greatly in setting up a project.

Budgets

The importance of time budgets has already been mentioned. Because time is a consultancy's primary product much effort goes into planning and recording its use. This can mean that the budgeting of rechargeables and disbursements can be neglected. It is vital to prepare comprehensive budgets or budget estimates for all project costs and (see chapter 12) to make the client aware of these costs. Irrecoverable cost/expense overspends can have a dramatic impact on cash flow and profitability.

Measuring and controlling

Even a small consultancy has a surprising number of costs and expenses flowing out. To assess the financial performance of the business these costs must be captured and allocated to projects or to administration and marketing.

Job numbers and job sheets

The easiest way to go about this allocation is to assign every project and every non-project activity a job number. As costs are incurred, they are coded to the appropriate activity. A cumulative visible record of financial progress is needed for each project. This is the job sheet or job report which can be produced by hand or by computer.

The job sheet should be up-dated weekly, and record the time spent and costs and expenses incurred on the project that week and cumulatively. The sheet should also show the total time and expenses budgets available, together with an estimate of the proportion of work completed.

Getting this estimate is usually the toughest data collection exercise, but it is vital in giving early warning of potential over-spends and projects which are going out of control: if only 50 hours have been spent on a project against a budget of 100 hours, the temptation is to skip on and assume all is well, which it may well be. On the other hand, if those 50 hours have resulted in only a quarter of the project being completed, it is time to start worrying and look for the problem that is leading to so many hours generating so little progress.

The job sheet system is very versatile and capable of almost infinite extension. But disciplined recording of time and expenses does not come naturally, so the system should be kept as simple as possible while still giving useful management information.

Expenses and order books

For job sheets to be useful they must give a timely and accurate picture of project costs. Expenses and bought-ins have a disconcerting habit of falling through the monitoring net because of the delay between placing an order and receiving an invoice from the supplier. This delay can be several months, by which time the job may be finished and the client have been billed. Going back and asking for further expenses is both unprofessional and frequently unsuccessful.

To avoid these problems an order book system is needed and should be run according to three golden rules:

- it should be made clear who can and cannot place orders with suppliers
- orders must only be made using official order forms, which should be numbered sequentially and kept by an appointed member of the studio staff
- no order should be issued without a signature from an authorized member of staff, a job number and an estimated price.

This regime should be applied to all expenses from cabs to Concorde, with dire consequences strictly implemented for offenders.

Each week the order book is analysed and the expenses incurred summarized by job number and shown on the appropriate job sheets. Thus project managers do not have to wait until an invoice arrives to see the true costs on their projects.

Recording time

Recording and analysing time use is the basic control mechanism in a design business. Even the smallest studio should adopt a simple but comprehensive system. The time sheet is at the heart of time analysis and every member of staff whose hours are chargeable should complete one.

The sheet should show the hours spent on each project, identified by job number, job description and client name. Hours allocated to holidays, sickness, new business, training and administration should also be shown, together with total hours for the week. Time sheets must be collected weekly and their information transferred to job sheets promptly.

Who gets what information?

Having established the means of monitoring projects, who should get what information? Designers should normally know the hours budgeted for their contribution to the project, and the project manager should receive the job sheet(s) for his or her job(s). Beyond this, the designers should also be encouraged to show an interest in the overall progress of a job, and project managers should be seen to use the information they request, otherwise the whole project control system will be considered a wasteful infringement on designers' time.

It is useful to have a project summary report which gives the following information for all active projects; client, project name, job number, project manager, time budget, time spent, fee budget, fees billed, expenses/costs budget, expenses/costs billed. This should be reviewed by the principals weekly and circulated to project managers.

Project reviews

Reviews should be both formal and informal, giving designers a sounding board for their ideas and reviewers the opportunity to guide and suggest rather than direct (see chapter 7).

Reviews give those running the project the opportunity to understand the designer's approach and ensure that things are not going wildly off track. This can be frustrating for the reviewer if a better, quicker solution seems obvious, but junior staff should be given the maximum opportunity to discover for themselves under guidance. The temptation to direct should be resisted until there really is no alternative.

Checking against the brief

Projects gather an internal momentum which can sometimes lead them away from the brief. During reviews the designers should be encouraged to explain their thinking in terms of the brief, rather than simply explaining what they have done. To do this, both designer and reviewer must be fully conversant with the brief.

Establishing direction

Early reviews should be used to agree direction for the project, even though many designers like to wait for as long as possible because they can always think of further approaches and committing to one or two is an act of intellectual courage.

Controlling quality

Although a large part of the review process is aimed at making designers more thoughtful and articulate about their work, an equally important objective is to ensure that quality standards are being met.

It is no use holding a major review the day before presenting to a client. If the process is to have any value and not be unbearably stressful, the reviewer must be able to offer constructive ideas with time available to improve the work if necessary. The review should be used as a dress rehearsal, in which the reviewer tries to think and question from a client standpoint.

Concept presentations

The objective of the concept presentation is to give the client a thorough grasp of the designers' thinking, and of their approach to the brief, as well as a broad-brush picture of how the concept, or a number of possible concepts, might look and feel. There must be enough meat to enable the client to chose which concept(s) should be taken forward to detailed design.

A presentation is a selling occasion, and the first real opportunity for a client to assess the consultancy's performance. As such it should be meticulously planned to show the quality of the designers' thinking to best advantage.

Who should attend?

Because a successful concept presentation is so important the audience must be carefully composed and prepared. Consultants must impress the purpose and objectives of the meeting upon their clients to ensure that:

- real decision-makers are present
- potential project 'saboteurs' are involved
- all the attendees are aware of the brief
- attendees understand the objectives of a concept presentation
- attendees know the presentation standard, eg they don't expect a colour video when they will actually get sketches and flip charts.

The presentation should always be made by the design team, and any attempts by the client to take away a packaged presentation which can be shown internally without designers present should be resisted. The direct client feedback received at concept presentations is invaluable to the development of the project.

Moving things along

Allow plenty of opportunity for people new to the design process to ask questions, understand the designers' logic and equip themselves to make a decision. The presentation should be orchestrated to favour the option(s) felt to offer the best solution. Designers must be prepared to champion that solution cogently. If a decision is not reached in the meeting then a deadline for a decision should be agreed. The team should be prepared to do additional work and re-present concepts quickly. When a decision has been taken this should be confirmed in writing with the client.

Handling disagreement

Sometimes a client is not happy with any of the concepts presented. In this delicate situation it is important to establish a conciliatory atmosphere and probe for the detailed reasoning behind rejection. The fact that both parties share the common objective of a successful and cost-effective design solution should be emphasized.

Perhaps the brief has changed without the client making this explicit. Here the consultants will have to make a judgement about the scale of the change and decide whether a request for additional fees, although unquestionably justified, is advisable.

A situation where the client just dislikes the concepts on aesthetic grounds is tricky; it can be argued that one subjective judgement is as good as another. Both sides should relate their views to the brief, and if there is deadlock the consultant must choose whether to make a stand, ask for further fees, or go back to the drawing board.

Teams sometimes lose sight of the brief. This should be spotted during reviews, but sometimes it takes the client to bring things back to earth. The designers must then go back, re-absorb the brief and start again without any additional fees.

Design development and detailed design

During detailed design, conceptual ideas must be refined into a series of specific applications resulting in designs which are sufficiently detailed for the generation of artwork, working drawings or engineering drawings.

Final presentation

Detailed designs should be presented to the client in a form that as nearly as possible simulates the appearance of the design as it will be produced or executed. By this stage, there should be few surprises in what the client sees. Detailed notes should be taken of what is agreed, including any changes or amendments.

Design freeze

Once the detailed design has proceeded to a point where, depending on the nature of the project, form, colours, materials, finishes and layout have been agreed between client and designer, the design needs to be formally frozen. The client should understand that any subsequent changes will cost money and may delay completion.

Changes and amendments

Clients always request minor alterations to the design. If the request is verbal, the designer should confirm the change to the client in writing, advising of any additional costs and delays that may arise. Significant changes should always be confirmed by the client in writing before work proceeds.

After the design freeze the recording and specific approval of amendments becomes doubly important, because the cost implication is likely to be greater and new working drawings, artwork or contract variation instructions may need to be issued.

Drawing registers

Project amendments can cause chaos if a well-managed drawing register is not maintained, especially on interior and product design projects. This register should describe and number the drawings issued, record to whom they were issued and when; and indicate if and when a drawing has been amended or superseded.

Prototypes and pilots

The reason for building a pilot store or indeed a prototype product is to test its usability, appearance or consumer appeal, and ease of construction before finally committing substantial resources to a roll-out or tooling programme.

Evaluation and research

The most important aspect of this phase of a project is evaluation. It is important to have clear objectives for the prototype and well thought-through evaluation processes. It should be stressed to the client that the purpose of piloting or prototyping is not to prove that the design is 'right', but to learn what final refinements will make it easier to produce or work better in the market place.

However, designers must beware the tyranny of consumer research. For example, market research on prototype batches of 3M's Post-it notes in four US cities voted them a disaster. Fortunately a product champion ignored the research and every office is witness to the result.

Implementation

Given the breadth of design activity, implementation involves a whole range of specific activities, depending on whether a project is graphics, product or interiors led. However, during all implementation phases a design moves from the drafting pad into some tangible finished form, usually involving the use of technical and craft processes.

The designer's role

The designer's role in this translation process is primarily one of quality control. The design passes from the designer into the hands of knowledgeable specialists (printers, converters, tool-makers, contractors and so on) who will inevitably suggest ways of fitting the design to their plant, process or skills which may substantially reduce costs but could alter the integrity of the design.

The designer should be able to take the lead in suggesting appropriate and cost-effective implementation methods and be able to monitor the results constructively. This requires detailed knowledge of the implementation processes relevant to the designer's specialism, without which it will be difficult to have a productive dialogue with the implementors. This knowledge gap is a great source of client disappointment, and frequently mars otherwise excellent projects.

The client's role

Since the client is paying the costs of implementation, which are likely to be many times greater than the design fee, and takes over ownership of the project at this point, he or she will have a keen interest in the process. The division of responsibilities between client and consultancy should be clear. It is important for the designers to stay accessible and involved even if the budget did not provide for this, since it is during the translation from drawings to real objects that the client can lose confidence. Day-to-day technical and financial constraints or staff apathy can leave the project champion feeling exposed and lonely.

The client may be asked to arbitrate and take decisions, for example on costs, which could undermine the design concept. At this point the designer needs to be able to rely on the client's confidence in the project to ensure support. Good project management will underpin that confidence.

Sign-offs

Sign-offs are a protection for both client and designer and reduce the risk of disagreement. Designers should be thorough in ensuring that clients sign off all art-work or approve detailed drawings before any production or construction starts. The designer should ensure that at the completion of a project or project phase, the client formally recognizes its completion.

Particularly in construction projects the designers should try to ensure a role for themselves in signing off completed work as satisfactory. This strengthens quality assurance and protects the client.

Using other professionals

Designers should welcome other professionals into the implementation team rather than try to stretch into areas beyond their competence; for example, using a quantity surveyor on larger interior projects to price work, negotiate contracts, resolve disputes and sign-off payments to contractors.

Getting paid

Chapter 12 deals with the importance of checking the credit worthiness of potential clients and agreeing terms of payment at the outset. With this foundation in place, the next step to getting paid is timely and accurate billing.

Invoicing

Steps should be taken to ensure that invoices are not queried and payment delayed:

- invoices should be issued at the time or stage agreed in the terms of business
- invoices should cover only completed work unless agreement exists to the contrary
- specify clearly what work or expenses are covered giving sufficient detail for the client to identify what he or she is being asked to pay for
- ensure that invoices go to the right address and the right person, particularly in large multi-location organizations
- invoices sent to the client with a personal note can speed payment but also run the risk of not being passed to the accounts department
- for overseas clients, invoices should be issued in the currency specified in the proposal and VAT included only where appropriate
- statements should be sent monthly since many companies only pay on receipt of a statement.

Invoice chasing

All invoices should be chased if not paid within seven days of their due date. Queries must be resolved quickly and credit notes issued promptly if required. Although the accounts department may be holding up payment, talking directly to the client and discussing any problems may help to unblock things. Persistence is important. If work has been satisfactorily completed the client should pay for it. Partners or principals should chase personally if any significant payment problem arises.

Last-resort methods

If a client is simply refusing to pay or claiming that work is unsatisfactory or incomplete the consultancy is in difficult territory. This is where proper sign-offs can pay dividends: without them disputes can degenerate into, 'We say it is satisfactory/you say it isn't' arguments that are difficult to prove. Taking such disputes to court is high risk, time consuming and expensive, and it is important to decide early on whether to make a write-off or to fight.

If satisfactory completion can be proved, there is a small armoury of last-ditch tactics available:

- solicitors' letter
- court action
- winding-up order
- debt collection service
- sitting on the client's doorstep
- withholding copyright
- withholding drawings or artwork.

Specialist advice should be taken when adopting any of these tactics.

Post-project reviews

When a project is complete, it is invaluable to undertake a post-project review. The review should have three strands:

- how well is the design performing?
- how well was the designer/client relationship handled?
- how efficient was the design process on both client and consultant's sides?

The review can cement relationships and shows a rigour which clients will welcome. The client will benefit by learning how to become a more effective user of design.

PR

PR should always be a joint activity involving designer and client. Angles and copy should be agreed with the client in advance and endorse the co-ownership of a project's success. The client has the right to refuse consent to PR activity.

Managing the story

Once agreed, the story should be carefully managed. Media opportunities need to be prioritized since once a story has appeared it may be less attractive to other (perhaps more prestigious) outlets which prefer exclusivity.

The target audience for PR should be carefully identified, bearing in mind that while it is fairly easy to get coverage in the design press, this reaches relatively few potential clients.

Getting coverage in national business titles is tougher but reaches a very influential audience. Trade press can be a valuable and more accessible medium.

Any PR activity should be supported by good written and visual material in a variety of formats.

Confidentiality

Care should be taken to obtain client consent for the timing and medium of release and also for any competition entries. It is important to stick to the agreed story and not to disclose confidential information in the desire to give a fuller picture. Ensure that journalists are aware of any embargo dates before which a story is confidential.

Further reading

Burstein, David and Staslowski, Frank (1989) *Project Management for the Design Professional.* New York: Watson-Guptill.
Design Business Association (1992) *DBA Managers Guides.* London: Department of Trade and Industry.
Fisher, Roger and Ury, William (1989) *Getting to Yes.* London: Business Books.
Green, Ronald (1986) *The Architect's Guide to Running a Job.* London: Architectural Press.
Lock, Dennis (1988) *Project Management.* Aldershot: Gower.
Mott, Richard (1989) *Managing a Design Practice.* London: ADT Press.
Salisbury, Frank (1990) *Architects Handbook for Client Briefing.* London: Butterworth.

Simon Stern/Sharp Practice

'Management is the least known of our basic institutions. Even the people who work in the business often do not know what their management does and what it is supposed to be doing, how it acts and why, whether it does a good job or not.'

Peter Drucker
The Practice of Management

14 MANAGEMENT IN DESIGN PRACTICE

David Rivett

If Drucker's assertion (opposite) is true for business as a whole, it is doubly true for many design businesses. Design is still at a very early stage in its transformation from a craft activity to a serious business. Analogies are often drawn between design consultancies and advertising agencies, and indeed many of the human issues they face are similar. However, it is instructive to reflect that JWT and Lord and Thomas were well established by 1900, Young and Rubicam was founded in 1923 and Leo Burnett in 1935. Set against this perspective most design businesses are very young indeed and have little accumulated management experience on which to draw.

In addition, most design businesses are still run by their founders, who started life as designers and have, often with little help, evolved a management style out of necessity rather than by desire. Thus the industry is full of pathfinders casting about for the best way to manage their practices. Not surprisingly there are many styles and many interpretations of the management role. The approaches most likely to succeed will probably have the following in common:

- clarity of purpose
- clarity of objectives
- good communication
- respect for others
- decisive action.

Corporate and personal objectives

It is vital to recognize that personal and business objectives can differ. Designers must be honest with themselves about their own needs and clear about their priorities. So if there is a desire to grow to anything larger than a two- or three-person studio, more and more time will be required to develop client relationships, win new business, look after financial matters and deal with staff issues. This either means that the designer/founders must become facilitators and managers, or that non-designers must be brought into the practice. If growth continues both will probably be required, so it is important for designers to acknowledge that the inevitable accompaniment to successful growth will be less time spent designing, more delegation, more sharing of responsibilities, more supervision; in other words more management.

Designers and non-designers

It is characteristic of professions that there is a fundamental division between those who practise and those who 'merely' assist. This is generally unhealthy and frequently gives rise to inefficiency and hostility. In building a practice, the designer can help to break down this barrier by acknowledging that without the involvement of non-designers he or she would be confined to a more limited canvas.

Consideration of the main areas of management activity in a design consultancy shows that, as the practice grows, the need for people whose primary skill need not necessarily be designing will also grow in such areas as:

- creative direction
- analysis/planning/research
- new business/PR
- finance
- project management/implementation
- staff development
- computer systems.

The practice should be a team in which everyone understands, and is encouraged to respect, the functions played by others. Progress will be smoother and surer if designers recognize that even those without a design training can think as designers and make a valuable contribution to the design process just as non-designers must recognize that designers can also contribute greatly in areas outside design.

Managing, leading or both

The essential function of management is to maximize the medium- to long-term potential of a business. Every decision a manager takes must sooner or later yield some economic benefit for the business or it is a bad decision. In design this can be a difficult proposition to sustain, because of the fear of a trade-off between financial performance and quality of work.

This fear is understandable, and reflects the difficulty of getting the right balance between the need for short-term profits and the desire to invest for the future by delivering more or better work than has been paid for. It can be very easy to justify project over-spends, increased overheads, and all manner of other costs on the grounds of building for the future. Some loss-leaders can legitimately be seen in this light, but not many. A manager's first duty is to ensure the business survives and makes money, so that its owners and employees can enjoy the future.

Thus the manager must be both a strategist, thinking for the future, and a pro-active tactician, balancing long-term aspirations and the investment they demand with the need to ensure day-to-day profitability. While performing this difficult balancing act, the manager also has to ensure that people actually have fun. Design is fortunately one of those activities people enter because they find it attractive; it engenders high expectations for stimulation, fulfilment and enjoyment. Meeting these expectations is another major part of the manager's role.

Profits today

Looking first at short-term survival, four areas make priority claims on management attention:

- new business
- quality of work
- efficiency of work
- cash.

Chapter 10 deals in detail with securing new business. Whatever approaches are adopted, management must have a finger firmly on this pulse. It is too late to think about new business initiatives when there is a week's work left. Good practice management demands constant review of the length and quality of the order book. Without new business nothing is possible. It is, therefore, the single most important measure of the practice's health.

Creating a culture in which everyone in the organization is trained to recognize marketing opportunities and treat marketing as inseparable from their other activities is an effective way of keeping new business healthy and front of mind.

Work quality is a many faceted issue in which there is much scope for subjectivity. From the practice management point of view, client satisfaction is the most useful measure, since it is the best indicator of whether briefs are being met and clients likely to return and/or recommend the practice to others.

Designers frequently complain that clients do not recognize good or bad quality in design. In the few instances where this is true, it simply leaves the duty to produce good design in the hands of the designer, and that 'good' design must still satisfy the client, so client satisfaction remains a valid criterion for judgement. Post-project reviews are an excellent way of assessing client satisfaction (see chapter 13).

Design efficiency is important since if the promised programme cannot be completed satisfactorily, within the time and cost specified, the practice will lose money, and if this is repeated project after project the end is predictable. Where there appears to be serious inefficiency, such as projects always going over budget, it is important to look not just at the design process but also at the sort of business being secured and the fees charged. In a competitive field it is tempting to under-price and over-sell, particularly when starting up. This can lead to a vicious circle in which everyone is rushed off their feet, quality may be compromised and still there is never any money in the bank. Chapter 4 looks at financial principles, but cash in the bank is a good rough-and-ready indicator of financial health. More businesses go bust because of cash flow problems than because of poor profits. Thus managing the cash is a key daily management activity.

If a manager can write on the back of an envelope an accurate summary of the four measures discussed above, he or she probably has a very good grasp of the health of the practice. This is a worthwhile exercise to do at any time, including now.

Building for the future

To succeed in the future requires a clear idea of the practice's objectives; what types of design does it want to undertake, in what geographic areas, for what sorts of client? A strategy must then be formulated defining what must be done to secure the objectives.

The concept of SWOT analysis comes from strategic planning: it is very useful in helping to define objectives and develop a strategy, and may be applied to people as well as organizations (see chapter 7). The process involves an honest appraisal of an organization's strengths, weaknesses, opportunities and threats. It is important not to be introspective when performing the analysis, but to concentrate on the market context and to recognize that weaknesses can also be opportunities.

Page 142 shows a simplified example of SWOT applied to a design consultancy. This sort of analysis is immensely useful in giving clarity to plans for the future, and in encouraging anticipation of pitfalls and set-backs.

SWOT analysis of a design consultancy	
Strengths	**Weaknesses**
▪ creative reputation	▪ poor project management
▪ language skills within team	▪ financial resources
▪ packaging expertise	▪ partners over-stretched
Opportunities	**Threats**
▪ European packaging	▪ stronger European groups/ ad agencies
▪ improve project management	▪ falling packaging margins
▪ introduce packaging technology	▪ increased resources needed to pitch
▪ expertise	
▪ new investor/partner	

The analysis also highlights the resources needed to pursue a particular path and compete successfully. To grow or recruit good people, provide technical support and good working conditions, to fund overseas new business initiatives, requires money. This must be recognized and the temptation to try to do too many things at once resisted. The costs of any growth initiative should be budgeted and the source of money to fund it identified at the outset. Over the long term, these funds will be in proportion to the profitability of the business. The economic performance of a design business can never be neglected; profitability is the freedom to shape one's own future.

Leadership

No business was ever just managed to success; somewhere along the line there was a leader who, whether loved or feared, had a clear vision of the future and his or her role in it. Whether they like to admit it or not, employees of any organization are looking for leadership. Instead of simply being told what to do, employees in a well-led company respond to evidence that those managing have high standards, offer clear direction, demonstrate good anticipation of the future, accept responsibility, act decisively and praise achievement. Above all, leaders must want to be in charge and accept that the buck stops with them.

Distinguishing management from administration

At the opposite end of the spectrum from leadership is administration. As with leadership, every business must have it, but the routine administration of a practice which, in boom times, can pass for management is not sufficient to ensure success. It is important to run constant checks that the practice is not on auto-pilot. Are assumptions queried, are actions pro-active rather than re-active, is client feedback actively sought, is the future constantly probed for opportunities and upsets?

Culture

An essential part of the leadership role is the establishment of a prevailing culture or personality for the practice. A culture is unlikely to be created in one big bang, but more by a whole series of implicit or explicit expectations; for example, clients will always be greeted with an offer of coffee; reception will always display foreign newspapers; desks will always be cleared at night; telephone calls will be returned immediately; no-one will travel first class; everyone should have a Mac; designers will always present to clients; and the practice will pay for a party once a month.

Over time, the fact that certain principles are visibly important will be understood by everyone involved, and will build a powerful culture. It is vital to choose carefully those things on which great store is placed. A focus on financial performance will shape a particular kind of culture; client service as a focus will grow a different culture; new business, another and so forth. To succeed, every business should have a set of core obsessions which are universally understood and constantly reinforced.

Communication

One of the most important aspects of a practice's culture is its attitude to communication. Frequent open internal communication about everything, from the level of new business and the long-term strategy of the practice to staff morale and the state of the premises, will help to involve people and encourage them to make their views known.

Managers who communicate clearly and honestly (and this does not necessarily mean slickly) will engender confidence and reinforce their leadership. As with all management principles there are some words of caution:

- having communicated the intention to do something, whether introducing a profit share to re-decorating the studios, action must follow the words
- having sought people's input to decision-making processes, that input must be seen to have influence
- messages communicated formally should be consistent with other informal and behavioural communications which help shape the culture
- the level of communication should be consistent, for example, if it is customary to share financial performance with staff this must continue in bad times as well as good or people will fear the worst. Consistency also requires that principals must have thought through their response to difficult issues.

Structures for the 1990s: managing growth

The simple answer to the question 'Why grow?' must be that since organizations need change to stay alive, at any instant they will be either shrinking or growing.

On balance, growth is to be preferred since it implies new challenges, bigger projects and room for staff to develop. But growth should not become an end in itself. Instead it should be the natural outcome of doing good work, having happy clients and wanting to explore new territory, both geographically and intellectually.

Specialist or multidisciplined

Many consultancies grew large in the 1980s by expanding rapidly across several design areas to become multidisciplinary. Some would argue that this was the result of a desire for growth for its own sake rather than a response to the needs of clients. In many instances the cost and difficulty of transferring a reputation made in one area of design to others was greatly under-estimated. Research indicates that clients give specialist skills higher priority than the ability to go one-stop shopping for design. Those managing the practices of the 1990s can benefit from the lessons of the 1980s, which suggest that leadership in a narrowly-defined sector, based on technical skill and depth of expertise, may be a more appropriate and less risky strategy for the current decade.

The overseas dimension

If a re-emphasis on specialization is desirable, where is growth to come from? In part from economic growth and increased appreciation of the skilled specialist designer's contribution to clients' success, but also from greater participation in overseas markets. This topic is going to be of fundamental importance to design consultancy

Key stages in the growth of a design business

Features	Management/organization	Comments
STAGE 1 **Sole practitioner** **Fee range** **£0-£150K**	■ designer carries out all functions ■ bookkeeper/accountant used on a freelance/consultancy basis ■ unless registered as a limited company, business and personal finances bound up together ■ VAT registration required when turnover exceeds £35K p.a.	■ business growth limited to capacity of the individual ■ projects limited to the skills/experience of the individual; unless undertaken with other designers ■ designer's personal time management must be highly disciplined
STAGE 2 **Small employer: up to 5 employees** **Fee range** **£50K-£500K**	■ founder joined by additional designers and possibly specialist staff (eg for new business or practice administration) ■ more administration required ■ with full-time staff to keep busy, constant effort needed on new business ■ change of legal status to limited company may be effected	■ larger projects become feasible ■ founder takes on serious legal responsibilities once staff are employed ■ founder needs to spend time managing other people and learning to delegate ■ business may be able to expand into new areas depending on skills and experience of staff
STAGE 3 **Medium-sized business: up to 20 employees** **Fee range** **£300K-£2m**	■ founder spreads management responsibility ■ greater sophistication required in financial control and reporting structures ■ designers split into teams servicing different types of work ■ medium-term (ie 6-12 months) budgeting and capacity management become important ■ dedicated and active new business operation required to keep full-time staff busy	■ founder must accept need to share management responsibilities among other partners/directors ■ founder may have to give up designing and concentrate on new business and client service and the ongoing development of business ■ larger premises and associated overheads may be unavoidable ■ outside finance may be needed from a bank or another investor to fund expansion; a business plan will be required
STAGE 4 **Business with 75+ employees** **Fee range** **£1.5-£6m**	**Stage 3 plus:** ■ sophisticated information and communications systems required to avoid information overload ■ planned staff development and appraisal systems needed ■ professional (and full-time) financial director and managing director likely to be required	**Stage 3 plus:** ■ need to guard against development of bureaucracy and creation of barriers between managers and designers ■ care needed to avoid loss of experienced staff who feel excluded from the management group ■ founders need to give thought to management succession

in the UK during the coming years, and chapter 16 deals with the international market for design and its particular opportunities and problems.

Organizational milestones

As they grow, design practices pass a number of milestones which signal the need to change structures and manage differently. The biggest of these comes when moving from being a sole practitioner to an employer, when suddenly the founder becomes responsible to and for others, and must communicate, share and delegate, skills which don't necessarily come naturally.

Working hard at delegation, and resisting the temptation to do more and more things oneself (believing rightly or wrongly that they will be done better) is one of the most important disciplines of a good manager. Only if employees are allowed to make mistakes and learn will a team capable of supporting growth be created.

By the time a practice reaches four or five people it will need simple systems for project management and financial control. Once it reaches 20 employees, internal communication and studio management will become important concerns, as will the size of order book needed to keep 20 people busy. Even two founding partners may be spread quite thinly in this size of practice, and delegation will need to be pursued rigorously, with principals deciding where they should focus their energies and when they may need to recruit specialist help.

When a business passes a turnover of around £3 million, or more than 50 employees, the need to summarize information and keep track of data and projects may require new management information systems and more formalized internal communication. To preserve the balance of expertise and experience the founders may consider how to bring additional people into the 'principal' group.

Each practice will have different milestones, but it is always important to be aware of the existence of step functions in growth where systems, structures and roles must be reviewed, and possibly changed to support the next stage of growth.

The partnership spirit

Although some large design businesses have been built as public companies, far more very large consulting and service firms in all areas of business have grown as partnerships. This does not necessarily mean they are legal partnerships (see chapter 3) but that they embody the 'partnership spirit'. At the core of the partnership spirit is an open-door policy at the very top of the firm which allows anyone of sufficient ability to become a principal in the business, with a say in its direction and a share in its financial success.

In any business dependent on intelligent self-willed people, 'going it alone' will always be an attraction and this should be recognized. However, the loss of experienced staff, and their subsequent replacement, is disruptive, time-consuming and expensive, and something which any manager should try to minimize by building an appropriate business structure and management style.

The strength of the partnership spirit is that it gives people an opportunity for self-determination and influence without having to leave the practice and strike out alone.

The shamrock organization

The shamrock organization is a term coined by Professor Charles Handy to characterize the type of flexible structure which may prove well suited to a decade of change and discontinuity. The three leaves of the shamrock represent three types of participant in the activity of a business. The first is the essential professional core who, by their skills and experience, differentiate a business from its competitors. This group is the only truly permanent part of the shamrock. Much routine work is undertaken by the other two leaves; part-timers and specialist sub-contractors (or freelancers).

For a design practice, the lesson is to internalize, and to staff-up for an activity only when using sub-contractors or part-time employees puts the practice at a competitive disadvantage. Consultancies should always be looking at ways to minimize overheads, since this is the best cushion against a down-turn in business.

The network

Networking has long been seen as a good way for individuals to get things done, and it is also effective for businesses. In a world of increasing specialization and internationalization, the ability of design practices to supplement their own skill base or operate overseas by networking will be increasingly important.

When considering networking alliances, it is easy to get carried away by enthusiasm and ignore potentially fatal flaws in the structure. For instance, does a superficial complementarity and a shared professional language actually obscure considerable overlap in long-term ambitions which could lead to competition and a reluctance to share clients? In international alliances, are there sufficient language skills on both sides to ensure effective business communication as opposed to dinner table pleasantries? If both parties share similar skills, even if located in different countries, who will take the creative lead; will one feel the inferior 'implementation' partner?

Alliances which work best seem to be between practices which have different skills and by combining can win projects that alone they would not be able to tackle.

Succession

As many of the first generation of UK practice heads reach middle-age, thoughts often turn to the continuation of the business should the founders want to reduce their involvement or retire. Successors are needed, and ideally they should come from within the practice, but this will only happen if efforts have been made over a considerable time to grow the second generation. This means:

- senior management recruiting people as good as or better than themselves
- embodying the partnership spirit in the business allowing junior staff room to make mistakes and learn
- adopting standards which everyone knows and understands
- investing in training
- conducting regular and structured staff appraisals
- giving as many people as possible exposure to clients
- involving potential successors in decision-making.

Introducing new blood

Over time, every team becomes tired and stale if it does not have an infusion of new members. Introducing new blood is a vital management task, all the more difficult because principals may need to recruit people who will be a threat to their often cosy group. This process needs great care if the chemistry is not to go badly wrong, and requires discussion with everyone who will be affected. Ideally such incomers should be positioned as having different and complementary skills to the existing team, or be at a recognizably higher level so that there is no justifiable sense of anyone being passed over.

Managing an exit

Perhaps the supreme management task is handling one's own departure from a business. This is the ultimate act of delegation and needs careful preparation. Too many design businesses lose their way once the founders sell-out or lose their enthusiasm. Growing successors is clearly a major part of the preparation. In particular the day-to-day running of the business should have been completely delegated before contemplating retirement.

Founders should move to a consultative role, perhaps on creative direction or new business. They should progressively limit project involvement and actively endorse the skills of others in the practice. The succession plan should be made clear to clients and staff, with the founders' lessening involvement being presented as a positive process leaving the high ground to others.

These are necessarily general guidelines, but the specific objective is always to ensure that managers or founders do not outstay their usefulness to the business. There is nothing more demotivating to staff than seeing people hanging on to management control when they have neither the ideas nor the energy to keep moving a business forward. On the other hand, a founder group which is seen to be actively managing its succession and moving up strong talents to take over can give enormous encouragement to younger people in the practice.

A design practice which can manage the transition to second-generation principals is one which has truly proved its maturity.

Further reading

Deal, Terrence E and Kennedy, Allen A (1988) *Corporate Cultures: Rites and Rituals of Corporate Life.* London: Penguin.

Drucker, Peter F (1989) *The Practice of Management.* Oxford: Heinemann.

Handy, Charles (1990) *Inside Organization.* London: BBC Books.

Handy, Charles (1991) *The Age of Unreason.* London: Business Books.

Hunt, John (1986) *Managing People at Work: A Manager's Guide to Behaviour in Organisations.* Maidenhead: McGraw-Hill.

Linton, Ian (1988) *The Business of Design.* Wokingham: Van Nostrand Rheinhold.

Mott, Richard (1989) *Managing a Design Practice.* London: ADT Press.

Porter, Michael E (1981) *Competitive Strategy.* London: Collier Macmillan.

Roberts, Wess (1990) *Leadership Secrets of Attila the Hun.* London: Bantam.

Lo Cole/Sharp Practice

'Not a word was spoken between us.
There was little risk involved.
Everything up to that point
had been left unresolved.'

Bob Dylan

Shelter from the Storm

15 DESIGN AND THE LAW

Henry Lydiate

This chapter is about the law relating to intellectual property. Although this is one of the most complicated areas of legislation affecting design, it is one which no successful design practice can afford to ignore.

Parliament and the courts in the UK have created the law governing intellectual property. Most of it is contained in the Copyright, Designs and Patents Act 1988 which replaced the Copyright Act 1956 (preserving some of the earlier Act's rules relating to work made before the new Act came into effect). The law relating to passing off and confidentiality derives from cases decided in the higher courts, which developed legal precedents for the protection of businesses against unfair competition.

Before a detailed review of intellectual property law, we must mention the law of contract and its vital role in the protection of designers' rights. The law automatically gives designers protection from abuses and freedom for self-exploitation, but leaves them free to make their own arrangements with clients and employees for clarifying and recording ownership and use of their rights. Written terms and conditions of contract or agreement in all commercial dealings are essential. Most legal disputes arise through the lack of documentation recording initial pitches/presentations, negotiations, eventual agreements, and subsequent agreed changes. Chapter 12 covers contractual relations in more detail.

Copyright

Design work made since 1 August 1989 is given copyright protection by the Copyright, Designs and Patents Act 1988, which came into effect on that date. Work made between 1957 and July 1989 is still protected by the new Act, which carries forward some of the old law governing such work. This makes it confusing, especially for designers whose senior colleagues may not have up-dated themselves. Where the old law is still relevant, both pre- and post-1989 positions are explained.

Protected works

All 'artistic works' are protected against unauthorized reproduction, publication, televising, and commercial dealings. The copyright owner's exclusive right to prevent such abuses arises automatically in UK law, as soon as a mark is made by the author. If there is no physical manifestation created by the author, there can be no stealing of that image and so no breach of copyright. Ideas are not capable of being protected by copyright law (but see the law on confidentiality, page 159). All hands-on design work is capable of being protected, because 'artistic' in copyright law means paintings, prints, drawings, photographs, sculpture, craftwork and architectural drawings.

To qualify for protection, design work must be original. It must not be copied from another work, so that its form does not originate from another author – even someone anonymous or long-since dead. Original skill and labour are essential; the important point is that there is original expression, not simply original thought.

Having qualified as original and artistic, the work must also have been made by a designer who qualifies for copyright protection. This means that when the work was made, the designer was: a British subject, a citizen of the Republic of Ireland, a UK registered company, or a citizen/resident/company in a country that is a signatory to the major international copyright conventions (see below). This qualification is easily achieved and is significant for three reasons. It illustrates that works have to be made in order to be protected; it shows that copyright law is only domestic (applying at home), and needs international treaties to ensure that other countries pass similar laws to give protection abroad to UK-based designers' works abroad; and finally, it illustrates the automatic nature of copyright, which arises the instant a qualified person makes an original visual work.

UK law has no registration or publication scheme, which is a common prerequisite for copyright protection abroad. UK-based designers will be protected in the international copyright convention countries automatically, so long as their work carries the international © symbol, their name and year of authorship.

International copyright conventions: main countries

Algeria	Ghana	Niger
Andorra	Greece	Nigeria
Argentina	Guatamala	Norway
Australia	Haiti	Pakistan
Austria	Holy See	Panama
Bahamas	Hungary	Paraguay
Bangladesh	Iceland	Peru
Belgium	India	Phillipines
Benin	Ireland, Republic of	Poland
Brazil	Israel	Portugal
Bulgaria	Italy	Romania
Cameroon	Ivory Coast	Senegal
Canada	Japan	South Africa
Central African Empire	Kenya	Spain
Chad	Laos	Sri Lanka
Chile	Lebanon	Surinam
Colombia	Liberia	Sweden
Congo	Libyan Arab Jamahiriya	Switzerland
Costa Rica	Liechtenstein	Thailand
Cuba	Luxembourg	Togo
Cyprus	Malagasy	Tunisia
Czechoslovakia	Malawi	Turkey
Democratic Kampuchea	Mali	United Kingdom
Denmark	Malta	United States of America
Ecuador	Mauritania	Upper Volta
El Salvador	Mauritius	Uruguay
Egypt	Mexico	USSR
Fiji	Monaco	Venezuala
Finland	Morocco	Yugoslavia
France	Netherlands	Zaire
Gabon	New Zealand	Zambia
Germany	Nicaragua	

Ownership of protection

Works made before 1989 are still governed by the 1956 Act rules for the purposes of deciding copyright ownership. Generally, the first automatic owner of copyright was the maker of uncommissioned pre-1989 works. But if such work was made by an employee as part of his or her duties, the first automatic owner of copyright was the employer, unless the contract of employment specified otherwise. The first automatic owner of copyright was the client in relation to most commissioned works. But, the commissioning client and design consultancy were free to make different arrangements so long as these were specified in the written terms and conditions of their agreement. Uncommissioned photographs taken pre-1989 were a curious exception. The first automatic owner of copyright was the person who owned the material on which the image was made, that is the owner of the negative film or photo-sensitive screen/plate and not the taker or maker.

Works made post-1989 are governed by the 1988 Act's rules for the purposes of deciding first automatic copyright ownership (usually the maker). The exception is the employee, unless the contract of employment specifies otherwise. Commissioning clients and design consultancies are still left free to make their own arrangements as part of their written terms and conditions of agreement. If they fail to do so, the maker will be the first automatic owner (or employer, if made by an employee).

After 1989 the maker of a photograph is the person responsible for making all the arrangements necessary for its creation, eg the artistic director of the shoot, not necessarily the person pressing the shutter. Computer-generated works are also included in the new copyright scheme, the maker being the person who made the arrangements necessary for the generation.

Clearly, the changes post-1989 now put all visual art and design practitioners in a much stronger negotiating position in relation to would-be commissioning clients, especially in the design field, who will inevitably want to buy the right to roll-out, reproduce, publish or otherwise use the original design work. Whether they will have to pay more for the right to do so can now be a legitimate matter for negotiation and agreement, instead of the unexpressed assumption it frequently used to be; silence on the subject is not to a commissioner's advantage.

Length of protection

Works made before 1989 were usually protected for the author's life plus 50 years after death, even where the author was an employee whose employer owned the copyright. Exceptions were published photographs and prints, protected for 50 years from the year of publication; unpublished ones could have protection forever.

Works made after 1989 are protected for the author's life plus 50 years after death, including all photographs and prints, even unpublished ones made pre-1989.

Transfer of protection

Owners of copyright cannot legally part with their intellectual property unless they execute a document signed by them giving or selling copyright to some other person or body. Contractual negotiations involving creation and future use of such rights need clear written details to protect both paying client and designer.

Licence to roll-out

Licences are legal jargon for permissions. The owner of copyright has the exclusive right to prevent and to authorize reproduction/publication/televising/commercial dealings. Only copyright owners can allow such uses of their works by giving licences. Although the law does not require such licences to be written, no well-run design consultancy authorizes use of its copyright work without prior written authority being given. Contractual negotiations with commissioning clients should include discussion and written agreement on this subject.

Breaches of rights

Anyone reproducing, publishing, televising or commercially dealing with a copyright work without the permission of the copyright owner, breaches the law. The copyright owner can bring civil proceedings to stop any anticipated infringement and receive compensation for past ones.

The following acts are permitted without the express licence of the copyright owner:
- research or private study
- advertisements of artistic works for sale
- criticism or review with acknowledgement
- news reporting of current events
- incidental use in film or television
- incidental use in another 'artistic work' (useful in design, so long as care is taken to ensure that the 'borrowed' copyright image is used in a truly incidental manner)
- public administration
- education
- works of sculpture or craft permanently situated in a public place or in premises open to the public
- subsequent reproduction by the same author in a later work.

Designs which qualify for copyright protection as original artistic works, but which are licensed by the copyright owner to be made into articles to be used commercially or industrially, will lose copyright protection 25 years after such articles have been marketed. Such designs will, however, be specially protected for 10 years by the new design right (see page 153).

Remedies for breaches

A copyright owner whose legal rights have been or might be infringed may have a choice of proceedings to bring before the courts; civil or criminal.

Civil proceedings can be brought by:
- the first owner of copyright in the infringed work
- the second or subsequent owner of copyright in the infringed work, to whom copyright has been assigned
- the person to whom an exclusive licence has been granted by the copyright owner.

Civil remedies include:
- injunction before or after a breach
- payment of profits made by the infringer to the original designer
- delivery of infringing copies, artwork and equipment to the original designer
- payment by the infringer of legal costs incurred by the original designer.

Criminal proceedings can be brought by:
- anyone, whether or not the copyright owner, who has evidence of an infringing criminal act; not all breaches are criminal, but most commercial abuses are.

Criminal penalties include:
- an unlimited fine or up to two years in custody for serious offences
- in addition, the court may order payment of compensation by the convicted infringer to the victim, forfeiture and delivery to the victim of infringing articles (or their destruction), plus payment of all or part of the costs of the prosecution.

Summary

Design consultancies require a working knowledge and understanding of copyright law, especially the rules about origination, first ownership, transfer and licensing. Equipped in this way, it is then possible to make managerial and administrative arrangements which ensure that:
- commissioning clients know at the on-set of negotiations who will own the copyright in the original design work, and what permissions are being given for future merchandising purposes

- designers know that 'origination' involving the visual work of other artists or designers will deny copyright to themselves and, in relation to 'borrowed' imagery that is still protected by copyright, will infringe that copyright
- original works always carry the © by-line, author's name and date of creation, during studio research and development and on finished artwork and any merchandised articles
- employers of designers clarify in written terms and conditions of employment, ownership of copyright and arrangements for keeping in touch with designers.

It is vital that design consultancies obtain written evidence of origination/authorship/first copyright plus ownership/licences/terms and conditions of contracts of employment and of freelance services. With clear written explanations of in-house line-management authority, pitfalls and disputes can be avoided or at least minimized.

Design right

Design work made since 1 August 1989 is given a new form of protection by the Copyright, Designs and Patents Act 1988: design right. Another form of intellectual property, it is intended to protect original designs for goods, products and packages against commercial competition (without the need to prove a breach of copyright). This is a vital new legal area for all designers and their clients to understand, since commissioned designs for articles and products will give the client (not the designer) design right, and will reduce the designer's copyright to 25 years (from the initial period of life plus 50 years after death).

Protected works

Designs of a two-dimensional nature (whose shape or configuration is for an article to be produced commercially or industrially) and three-dimensional maquettes or prototypes of such an article, are automatically protected by design right. If the merchandising venture is on a small scale (up to 50 articles), then design right may not be available and the artist/designer/craftsperson must rely on copyright for protection against unauthorized reproduction and commercial dealings.

Like copyright, design right arises automatically in UK law, and requires no registration or other formality. It gives to the design right owner the exclusive control over any manufacture and distribution of the design products. In effect, it replaces copyright for the design right period. In respect of designs that may not have copyright protection, because they are mainly functional and only incidentally artistic (such as clothing), it is the best form of protection. As with copyright, the design must be original. This excludes designs for commonplace products, for 'must-fit, must-match' tools or equipment (spare car parts), and designs that use the same or similar designs originated by an earlier author (even someone anonymous or long-since dead). Surface decoration, interiors, structures, and purely graphic/typographic/photographic works, are all excluded; as are methods or principles of construction (inventions) which are patentable.

Having qualified as an original two- or three-dimensional design right work, protection will only be available if the designer was a qualified person when the work was made. As with copyright, this includes UK-based designers and design companies, who are protected automatically both at home and in the European Community countries. Unlike copyright, protection outside the EC is only available in countries whose designers are offered design right protection in the UK.

Ownership of protection

Design right is first owned by a commissioning client automatically, unless the original commission agreement specifies otherwise. It is vital, therefore, that designers negotiating deals for the creation of new articles discuss and agree who will own the design right, and who the copyright. The designer automatically acquires design right in uncommissioned work; employee designers do not acquire design right, which goes to their employers, unless the contract of employment states otherwise.

Length of protection

Works are protected for five years before the product is marketed, which is a confidential development period running in tandem with any copyright in the work. From the time articles are made available in the marketplace, the right runs for a further five years, then for a final five years when the design right owner must give licences to anyone willing to pay a licence fee (if the compulsory licence fee cannot be agreed, it is decided a copyright tribunal or by the civil courts).

Transfer of protection

Owners of design right wishing to sell or donate their right must do so in writing; it is essential that commissioning clients' contracts clarify the matter in any event.

Licence to roll-out

Everything described in the section under copyright applies to design right, especially where the designers have developed an original product without a commissioner, and are merely selling a licence to merchandise/roll-out.

Breaches of rights

The section dealing with breaches of copyright more or less applies in relation to breaches of design right; bluntly, unauthorized commercial exploitation, including import and export of infringing articles, is a breach during the 15-year period of protection; and unauthorized reproduction of the designs during the first five-year confidential period is likewise a breach.

Remedies for breaches

The civil remedies available for copyright breaches are generally available for design right breaches; criminal remedies are not.

Summary

Design consultancies need a working knowledge and understanding of design right law, in tandem with copyright law. Written evidence of origination and licences to use designs are essential. The matrix on page 160 shows how design right interfaces with copyright and other intellectual property rights.

Design registration

The Registered Designs Act 1949 created a scheme for the protection of decorative and ornamental designs. Although such designs may be protected by copyright, it is also possible to register them under the 1949 Act scheme. To qualify for registration, the designs must be previously unpublished in the UK, and be mainly aesthetic rather than functional. Accordingly, designs for products or articles whose shape is dictated solely or principally by their function, are excluded; as are graphic, photographic and typographic works designed for printing and publishing.

Protection lasts for five years after first registration, which can be renewed four times up to a maximum of 25 years (coinciding with the 25-year copyright period applicable to designs marketed commercially or industrially).

The right to register is acquired by the designer or employer (as with copyright) if the design is uncommissioned. With commissioned designs, the client acquires the right to register (unless the commissioning contract states otherwise). Transfers and licences of the registered right are possible, but they must also be registered at the Design Registry. Once registered, the design is given a registration number which is normally (though not required to be) marked on the product or article carrying the design; together with the words 'Registered Design'.

Registration can be an important safeguard, particularly when decorative designs are marketed abroad. UK registration gives the registered owner six months in which to register the design abroad, during which time it is still protected even though unregistered elsewhere. Typical designs which appear to benefit from registration are: textiles (themselves) or designs to be applied in or on textiles; patterns for coverings of floors and walls; and ornamentation (but not wall plaques or medals).

Moral rights

Copyright protects makers of original works of art/craft/design against unauthorized economic exploitation. In the UK until 1989, copyright legislation gave little or no protection to such makers against abuses of their reputations: public denial of their authorship, and derogatory physical treatment of works no longer in their possession, control or ownership. As from 1 August 1989, protection against such non-economic abuses was given by the 1988 Act: moral rights.

Protected works

Designs are protected against moral rights abuses, so long as they are original artistic works as defined by the Act's copyright provisions (see copyright on page 149).

Types of protection

Makers, including designers, are given three moral rights.

- ***The right to be identified as author*** whenever work is
  - published commercially
  - exhibited in public
  - televised (broadcast, cable or satellite)
  - included in a film shown in public or in copies of such a film issued to the public.

The identification must always be clear and reasonably prominent. It must be in a manner likely to catch the attention of any person acquiring a copy, seeing the exhibition or television programme; on a building, it must be visible to people entering or approaching (designs for structures give the architect the right to be identified as such). The right will be given virtually automatically; the only legal prerequisite is that makers have asserted the right, that is, included it in written terms of agreement with clients.

- ***The right to object to derogatory treatment*** whenever work is subjected to any
  - addition
  - deletion
  - alteration
  - adaptation.

Such treatment is derogatory if it distorts, mutilates or is otherwise prejudicial to the 'honour or reputation' of the maker. The right to object only applies to a work which has been treated in such a way, if it is

- published commercially

- exhibited in public
- included in a film shown in public or in copies of such a film issued to the public
- dealt with commercially.

The right is given automatically; there is no need for assertion.

• ***The right to prevent 'false attribution' of authorship*** whenever a designer is identified (expressly or impliedly) as author, in or on any work they did not make, they may prevent any commercial or public dealings with such works. Also, an original designer may prevent anyone commercially dealing with work they made, but which has been altered since they parted with possession of it; the same applies where copies are dealt with as if they were the unaltered originals, or as if the original designer had authorized the changes.

Exceptions from protection

The right to identification as author is not infringed:

- if the work is computer-generated
- where the author's employer acquired copyright and the employer has authorized the use of the work without insisting on the author's (employee's) identification
- in film/television news reporting of current events
- in film/television or in another artistic work (eg collage, montage, mixed media, and so on) so long as it is not the main image screened or represented
- by educational establishments in examination questions
- in public administration
- if the work is also protected by design right (eg a product or package)
- if the work is made, or is made available, specially for a newspaper/magazine/periodical/encyclopaedia/dictionary/year book/other collective work of reference
- if the right has been waived in writing (ie written and signed by the author giving up the right) or the author has consented
- if the design is for a typeface
- if the design is created by an employee as part of their duties
- if copyright in the work has expired.

The right to object to derogatory treatment is not infringed in all the circumstances described above in relation to identification, and if the work is:

- modified to avoid the commission of an offence (public disorder, indecency, obscenity, blasphemy, profanity) or to comply with a requirement of the law
- modified by the BBC to avoid public offence, encouraging or inciting crime or disorder (so long as there is a sufficient disclaimer in relation to the author's non-approval of the change).

Length of protection

Identification and derogatory treatment rights last as long as copyright (the lifetime of the author plus 50 years after death); false attribution rights last for the author's lifetime plus 20 years after death.

Transfer of protection

Moral rights cannot be transferred, but can be waived by the designer (so long as such waiver is in writing).

Remedies for breaches

Designers can bring proceedings in the civil courts in the UK and overseas (see page 22), for orders preventing infringements anticipated, stopping or correcting any occurring, financially compensating for damage caused and for legal costs.

Summary

As from 1 August 1989, any act done to work (whether or not it was made before that date) which infringes designers' moral rights, will be caught by the new legal provisions.

Design consultancies have four main concerns in relation to moral rights. First, to ensure that their management practices and procedures include clear written explanations of moral rights to in-house designers. Rights to identification, for example, will not apply (unless agreed contractually by the board of directors/partners in relation to any in-house designer); but integrity rights (not to suffer derogatory treatment) and false attribution rights may apply where, for example, the studio agrees to publish the designer's name as author of two-dimensional design work.

Having clarified moral rights in relation to in-house designers, consultancies also need to ensure that personnel who have been authorized to engage freelance sub-contractors (eg photographers, illustrators, architects), apply a code of practice in relation to the copyright and moral rights of such freelancers. In these situations, the studio will be the commissioner and the freelancer the author or rights owner.

Just as the contractual bargaining power of individual artists, designers, photographers and architects has been increased by copyright and moral rights law, so design consultancies themselves will carry new legal powers into negotiations with their would-be commissioning clients. Accordingly, having established codes of practice in relation to in-house designers and the commissioning of freelance sub-contractors, studios also need to create a third code of practice for dealings with would-be clients.

Some clients will only need to own copyright, others only design right; some will only need a copyright or design right licence. Most clients will never have heard of moral rights. The studio will need to explain, discuss, negotiate and agree such matters with clients. In particular, they should discuss:

- authors' credits (the studio itself, in-house designers, and freelancers with whom the studio has sub-contracted to give credit, eg a photo or illustration credit)
- quality control (by the studio or its freelancers over any processes of reproduction of original designs commissioned from or through the studio, so as to prevent derogatory treatment).

Photographic privacy right

A new right is created in the 1988 Act and is given to commissioners of private or domestic photographs; it is the right not to have:

- copies of the work issued to the public
- the work exhibited or shown in public
- the work televised.

The commissioner's privacy right is not infringed:

- in film/television or in another artistic work (eg collage/montage/mixed media, and so on) so long as it is not the main image screened or represented
- in public administration
- if the right has been waived in writing (ie written and signed by the commissioner giving up the right) or the commissioner has consented
- if copyright in the photograph has expired.

No assertion is necessary, but it is sensible for commissioners to say so at the time they engage the photographer. This right is particularly useful for studios commission-ing photographers to produce images which need to be kept confidential; copyright in the photographs may be irrelevant to the studio, but confidentiality may be essential.

Patents

The 1988 Act and the 1977 Patents Act together provide for the protection of inventions, and give a monopoly to registered inventors for a specified period of time. Products or methods of manufacture which are thought to be novel or to involve an inventive step, and which are capable of being applied industrially, may be registered and acquire patent protection. There are certain exclusions: mathematical theories, computer programmes, scientific theories and aesthetic or artistic works.

Protection lasts for four years initially, and is renewable each year for up to 20 years. Transfer of protection by the registered owner is possible, as are licences or permissions to manufacture or produce; they must be given in writing and registered with the Patent Office. Applications for patents can be expensive: they require written specifications and drawings to be filed, and then a 12-month period during which the invention is examined and tested against the statutory criteria for registration (or grant of the application), followed by a further examination as to its novelty and industrial applicability.

Patent agents are usually commissioned to advise and assist in the procedure; their fees can be high. An employee usually has no rights to apply for a patent to cover an invention created as part of his or her normal duties. However, if the patented invention is of substantial benefit to the employer, the employee inventor may have a right to be compensated by the employer. Uncommissioned designers are free to register or apply for the grant of a patent, as they wish.

Trade and service marks

Similar to patents, there is a scheme for registering symbols or words identifying a particular class or type of goods or services. The aim is to protect brand identity against competitors and achieve a monopoly for the name or logo in the marketplace. For designers, it is particularly important to understand this area of intellectual property law, so as to be able to advise and assist clients commissioning designs which might include new brand names, corporate names, new service titles, and/or graphic logos or marks.

Marks could, therefore, involve the name of the business or a version of it, a new name or word, even an invented one. It is not possible to register commonplace or accepted words such as place names, first or surnames, or the names of ordinary articles in everyday use.

Individuals, companies and partnerships may apply for registration. Applications are made to the Registrar of Trade Marks, by designers themselves or through Trade or Service Mark Agents. The Registrar checks for previous registration of the same or a similar mark, and considers questions of exclusion.

Registration is initially for seven years and can be renewed infinitely, upon payment of standard fees. There is no legal requirement for the goods or service mark to include ®/™/SM, but these symbols are usually used to give notice of registration. Registered marks may be transferred or licensed, but this must be done in writing and registered with the Registrar of Trade Marks.

Passing off

Designers often look to copyright law for a remedy when they, their products or businesses are imitated. This often happens in relation to products which become well known and are not exactly or even similarly reproduced, but are 'passed off' as the genuine article. Sometimes an article is not an imitation of an original, but is misrepresented as being the work of a well-established designer or studio, passing off such articles as originating from a more established and reputable designer or studio.

Other intellectual property law may well offer remedies against such passings-off, but frequently the imitations in question are so far removed from the aggrieved designer's originals that it would be difficult to establish a visual connection for copyright, design right or moral rights purposes. Moreover, where imitation services are provided as if the 'genuine article', copyright, design and moral rights are inappropriate (since they only cover hands-on design work).

In these circumstances, designers have to consider the common law remedy of passing-off proceedings in the civil courts. In order to succeed, the aggrieved designer has to prove:

- the imitator has represented to the public that designs/articles/services were originated by the aggrieved designer
- the aggrieved designer/consultancy had an already established reputation in the marketplace
- the imitator caused the marketplace to believe that the imitations were originated by the aggrieved designer.

In defence to such actions, it is frequently claimed that the imitator did not intend to deceive anyone: 'I'm sorry, I didn't mean it'. This is irrelevant; if the marketplace was deceived by the imitations, that is enough to establish passing off.

Confidentialities

Design consultancies often need to be able to prevent competitors or hostile businesses from using information or ideas confidential to the studio. For example, ideas, schemes or plans for design works, which have not been put down in a material form and so have not created a copyright, might be acquired and used by another designer to make their own work. The originator could not prevent such use by using copyright or design right law, but might be able to do so by using the common law remedy of 'breach of confidence'.

Any aggrieved designer has to prove the following, in order to succeed:

- the information used was confidential to the aggrieved designer
- the information used was communicated in circumstances which the receiver of it also realized were confidential
- the receiver made an unauthorized disclosure of that information without permission of the aggrieved designer.

Confidential information includes anything written or oral, which the aggrieved designer had spent labour or money creating, which was not trivial, and which was not common knowledge.

Confidential circumstances depends, literally, on the circumstances of each case. What was reasonable? For example: a designer submits written proposals, in confidence, to a would-be commissioner for a major and complex contract; the proposals describe the proposed work, but do not contain any physical images by way of maquettes, final drawings, models and so on. The commissioner then uses this proposal to commission another studio to execute the first designer's work. There was no copyright or design right breach because no designs were actually made, presented and copied; only the ideas and approach were lifted. An action for breach of confidence would be the appropriate legal remedy.

Design and the law: interfaces

Law	Length of protection	Goods Products Packages	Graphics Illustrations Drawings	Photo-graphic prints	Decoration Ornament-ation	Interior designs	Structural designs	Advice & assistance services	Typefaces	First owner of protection
Copyright: 50 repro-ductions or less	designer's life +70 yrs	√	√ no.repro-ductions irrelevant	√ no.repro-ductions irrelevant	√	√ no. rolled out irrelevant	√ no. constructed irrelevant	n/a	n/a	Originator unless an employee
Copyright 50 + repro-ductions	25 yrs	√	n/a	n/a	√	n/a	n/a	n/a	√	Originator unless an employee
Design right	up to 5 yrs before marketing then 10 yrs	√	√ for first 5 yrs as drawings	n/a	n/a	n/a	n/a	n/a	n/a	Commissioner; if uncommissioned, originator unless an employee
Design registration	25 yrs	√ only visual features, not forms	√ visual feat-ures of goods, prod-ucts, packs	n/a	√	n/a	n/a	n/a	n/a	Commissioner; if uncommissioned, originator unless an employee
Moral rights: identifica-tion + quality control	designer's life + 70 years	√	√	√	√	√	√	n/a	n/a	Originator unless an employee
Moral rights: false attributions	designer's life + 20 yrs	√	√	√	√	√	√	n/a	√	Originator unless an employee
Patent	up to 20 yrs (5x4)	√	n/a	n/a	n/a	n/a	n/a	n/a	n/a	Inventor unless an employee
Trade mark	infinite	√	√ logos & symbols	n/a	n/a	n/a	n/a	n/a	n/a	Whoever first registers
Service mark	infinite	n/a	√ logos & symbols	n/a	n/a	n/a	n/a	√	n/a	Whoever first registers
Passing off	infinite	√	√	√	√	√	√	√	√	Whoever first achieved a market
Confiden-tiality	infinite	√	√	√	√	√	√	√	√	Whoever gives confidential information

Unauthorized disclosure: most disclosures will be unauthorized, unless the user can show that it was in the interests of the public to receive that information, say to help with police enquiries.

A person aggrieved by a breach of confidence may sue for damages and/or an injunction to prevent future breaches; not only is the user liable for the breach, but also for anyone else who uses the information after the initial unlawful disclosure.

Best practices, which can avoid or mitigate actual or potential breaches of confidence, include:

- all conversations/phone calls/meetings being recorded in writing or on tape
- all correspondence or other documentation being expressed to be 'in confidence'
- all confidential meetings/presentations/pitches being confirmed in writing beforehand, stating that everything will be 'in confidence'
- all such meetings being confirmed after the event in writing, summarizing key disclosures made in confidence and confirming that they remain so
- not disclosing any information/ideas/studio practices/policies to anyone who might use such disclosures to their own advantage or against the studio's best interests
- ensuring that written terms and conditions of engagement of all in-house personnel include requirements not to disclose any information about the consultancy to outsiders, save in specified circumstances
- ensuring that contracts commissioning independent freelancers require them to keep confidential all information acquired about the consultancy, before, during and after the execution of the commission.

Conclusion

Keys to successful practice, including the avoidance of legal disputes, are:

- knowledge and understanding of the legal framework within which all designers must operate at home and abroad (see chapter 3)
- knowledge and understanding of 'design and the law' (this chapter)
- recording all conversations, meetings, negotiations, agreements and variations.

Designers may find the following tips useful for handling actual or potential commercial dealings:

- *hesitate* before agreeing to do or not to do anything on behalf of the studio
- *negotiate* using written checklists of items that must be discussed and agreed before any commitment is made
- *notate* through written or taped records of matters discussed and/or agreed
- *communicate* indelibly upon design work to warn the marketplace of ownership of copyright/moral rights/design right/trade and service marks/patents; and in writing to confirm the results of meetings/phonecalls/agreements or disagreements, 'in confidence' where relevant
- *duplicate* by keeping copies of all correspondence and photographic records of design work that has left the studio temporarily or permanently.

Whether practising at home or abroad, adhering to the suggestions and practices for use of the material contained in this chapter will do much to enhance the reputation of the consultancy and will also help to minimize or even avoid legal disputes and pitfalls that trap both the careless and the ignorant.

Claudio Munoz/Sharp Practice

'Business is complex; the world is changing rapidly; the organisation is a maze of complexity; how can it be simple? Successful companies are trying to cut through the noise, formulate a clear vision of the market and how to be positioned in it, eject cumbersome hierarchies and rigid managment planning and proceed on the basis that the only known quantity is change.'

Laura Mazur

Marketing 2000

16 THE INTERNATIONAL MARKET
Joe Tibbetts

Design can no longer be practised within a UK-only context. The world is an increasingly complex and immediate cultural matrix. It is becoming clear that design has more to offer in making the global culture complex accessible to humankind than any other communications discipline, science or technology.

On a more pragmatic note, the trend for design clients to operate on an international basis is both growing and accelerating. Design businesses which aspire to service international clients must embrace an international strategy and adopt an international mind-set. Even those design companies which aspire to work only in their local markets, and few will be able to afford this luxury, will be forced to understand and acknowledge the multicultural influences which are increasingly affecting their clients' markets.

It is tempting to view subjects such as cultural convergence and the rapid rate of development in communications technology as fascinating but dubious fantasies propounded by slightly crazed futurologists but the cultural, political and economic trends discussed in this chapter are already powerful factors in our everyday life. The nature and dynamics of the society we live in are changing at a remarkable rate and for the foreseeable future, change will be the only constant in our professional lives.

Trends and factors affecting the global design market

There are three dynamic trends and one key factor affecting the global market for design in the 1990s. These three trends are having, and will increasingly have, significant effects on the provision of design solutions for niche (generic), micro (local) and macro (multinational) markets. The three trends are:

- the trend towards international environmental interdependence
- the trend towards international economic interdependence
- cultural convergence.

The key factor driving all three trends is:

- the increasingly universal and rapid access to information made possible by modern communications technology (CT).

The trend towards international environmental interdependence

During the 1980s concern over the ecological impact of everything that humankind does increased, exacerbated by disasters like Chernobyl and the discovery of effects such as the hole in the ozone layer. At the beginning of the 1980s green issues were the concern of a so-called 'lunatic fringe'. By the beginning of the 1990s green issues were near the top of every political, economic and social agenda.

Green issues are global issues, and in addressing these issues designers are obliged to address them in a global context. Within the global ecological context, a parochial mind-set or isolationist attitude will render designers unfit to practise.

The trend towards international economic interdependence

Historically, national attitudes towards home and export markets have been, 'every nation for itself'. In the past nations have protected their own markets, and have adopted imperialistic tactics in the acquisition of other nations' markets. The 1970s saw the concentration of wealth in the hands of countries such as Germany, Japan and the OPEC Arab nations. This gave rise to economic traumas, such as the oil crises in the 1970s, and entrenched protectionism, characterized by the imposition of import quotas and obstructive import procedures.

By the 1980s, trading strength, based on balance of payments surpluses, had polarized to a few countries. This development provided energy for the globalization of supply, driven by the need for the strong economies to find new markets. In turn, the globalization effect highlighted the fact that the health of a global exports-based economy depends on the economic health of the markets which buy its exports.

Self-interest was (and still is) the driving force behind the growing acknowledgement of economic interdependence. The major players, in order to prosper and grow, have been forced to adopt their export markets as home markets. Long-term development and nurturing of those markets, rather than the pursuit of short- or medium-term profits, have become the prime objective.

More and more design clients are coming to view the world as one market. Increasingly the global market is divided into different sectors, defined by lifestyle and aspiration, rather than national or geographic boundaries. Seeing the world as one market with different sectors rather than a number of different markets is not, as might appear, just a matter of terminology. Market researchers and client companies are already using criteria which ignore borders, and define the cross-cultural consumer by attitude and emotional profile.

For reasons that will also become clear later in this chapter, design is the pre-eminent international commercial communications mechanism. Consequently, design has a pivotal role to play in the development and dynamics of the interdependent global market place. The nature of this role dictates that design and designers acknowledge the globalization factor in both the designs they produce, and the way they service their clients.

The trend towards cultural convergence

Cultural convergence means simply that there is less and less difference between life in one country and life in another. There are still, of course, many differences in the way people conduct their lives in different countries, and these will remain for the foreseeable future. The cultural convergence effect is already a market reality, and one which will accelerate in the next 20 years. Differences in national, and indeed local, culture are based almost entirely on the perceptions of the population. In reality we all 'need' the same things: food, air, freedom, social and sexual intercourse, education and so on; differences in national culture stem from the way we perceive our needs and how we prioritize them. In considering the key factor – communications technology – we can see that both the media mechanisms and the messages they transmit are globalizing rapidly and becoming ever more powerful. Never before have consumer perceptions been under such pressure to homogenize.

The three trends outlined above are essentially market forces. If the market did not demand otherwise, humankind would find ways, as we have done in the past, to ignore environmental imperatives. Similarly, political pride and party political agendas would ensure that the implications of economic interdependence were ignored, and national pride and chauvinism would obstruct cultural convergence. The power of the marketplace cannot be resisted.

The impact of modern communications technology

The international mass media, cinema, television, and popular music transmit messages which effectively control the cultural perceptions of the mass market. They are the true 'market makers' in both consumer perceptions and in the consumers' understanding of products, services and 'packaging'.

There are no effective barriers to the transmission of these culturally homogenizing messages. As Stephen Spender wrote in *Granta 30*, 'The Berlin Wall may have prevented East Berliners reaching the West, but it was leaped over and penetrated at a million points by TV and radio bringing East Berliners news and images of the lifestyle, vitality and competitiveness of the West'.

Before the invention of printing, culture converged at the speed of the horse. Religious and social missionaries travelled from settlement to settlement disseminating ideas. With the invention of printing, the rate of cultural convergence accelerated, the dissemination of ideas no longer required the presence of the messenger. In the twentieth century the process accelerated still further with the advent of cheap colour printing, broadcasting and telecommunications.

Since the 1970s, however, the global marketing of moving images, such as international block-buster movies, supra-national news coverage and global marketing of television programmes have wrought a significant change in both the nature and the degree of global media effectiveness.

Changes in culture and perceptions which proceed from the operations of the mass media now become apparent not as predictions but as *faits accompli.* At the moment the transmission takes place the consumer is informed not only that 'here is a new product or service which is covetable' but also that 'the rest of the market is assimilating this within the same time frame'. In the global marketplace, the new is only new for a very short time, and the present is colliding with the future.

CT is driving the trend towards globalization of the markets. Because of television, telephones, faxes, remote control of communication services, electronic mail, free access data bases and so on, everyone is aware of how their neighbours live and what their neighbours buy. Almost everyone wants a piece of their neighbours' quality of life and only peace, prosperity and ecological viability will allow that to happen.

Developments in CT are in many ways a paradigm for the globalization of the marketplace and vice versa. It is difficult to imagine the one existing without the other. The immense amounts of freely and speedily available information necessary to the globalization of the market could not happen without the revolution in CT. Conversely the development of global trading patterns and structures has produced a fertile market for CT products and services.

Communications technology and the design sector

Not only is CT a key factor in driving global market trends, but it also has significant implications for design companies and the way they operate within their markets. CT has a major part to play in working the market for design services. Only those companies which use CT to keep pace with the rapidly changing commercial environment will be able to achieve the flexibility necessary to address new markets.

- CT, properly used, is a powerful tool which can increase the geographical reach and spheres of influence of even the smallest design company.
- CT can enable companies to produce and deliver design solutions of the highest quality without over-extending human and corporate resources.

In a market where the client is ever more demanding, and where speed in the creation and delivery of a service may be the only factors differentiating one supplier from another, the implications and possibilities of CT cannot be ignored.

Design in the multinational marketplace

We said earlier that 'design is the pre-eminent international commercial communications mechanism'. This is neither vain boasting nor wishful thinking. There are two factors which will assure design of this central role in the development of the global market place:

- the Tower of Babel syndrome – the problems of a polyglot market
- information overkill.

The Tower of Babel syndrome

In the past, design has styled itself as an international language. This claim has sometimes been made without much justification, but there is now the opportunity, indeed the necessity, for design and designers to make good this claim.

Some important areas of activity, like international physics and international air traffic control, have adopted English as their official language. Despite this, the global market is a polyglot market and will remain so for the foreseeable future. In the next two decades simple commercial motivation will ensure that design clients address more markets which straddle linguistic boundaries. This will be the case in all areas of commercial activity, not just in the packaging of fmcgs.

All cross-cultural marketing relies upon commonly understood codes and icons. The need for such shared semiotic systems will continue to grow to the point where such systems are essential for the market and marketing to function. The marketing of financial services, for example, relies heavily upon packaging and explaining difficult financial concepts to an effectively semi-numerate population. The suppliers of financial services, such as insurance, investment, and banking, will be unable to resist the opportunities presented by foreign market blocks. They will increasingly rely upon design to explain their products to the market.

The marketing, merchandising and process of travel rely heavily upon internationally understood information systems. As the consumer adopts a more global mind-set, more people will visit more destinations and travel more frequently. Design will have to provide product branding, packaging, promotional material and transit facilities which are equally accessible to a New Yorker and a Turk.

Paradoxically, global consumers will all be foreigners in their own marketplace. Design and designers will be given the responsibility of making both niche markets and the marketplace as a whole intelligible to the consumer.

Information overkill

As markets globalize, information design specialists, both within in-house teams and as bought-in 'experts' will become a vital part of addressing any design problem in any design discipline. The nature of information design (ID) is widely misunderstood. Some designers think that ID is of concern specifically to their own discipline. Conversely, some designers think that information design is of concern to all disciplines except their own.

The reality is that information design is an integral part of all design. Packaging and promotional material, signage and site logic, passenger transit systems and facilities, operating logic and symbol layout for new technology, video display layout for interactive applications, maps, newspapers, the interface at pavement level between commerce and the consumer and so on are all dependent on information design in order to function.

Japan has been described as the most information-intensive society in the world. The information overkill, the amount of information that the ordinary consumer has to deal with in every aspect and location of daily life, far outstrips current experience in Europe or North America.

Japan, however, is only one country with 125 million people and only three major language divisions. Global markets will multiply and complicate these factors many times. As the frequency of cross-border communications increases, both in the form of physical movement and in technology-carried information, there is potential for information overkill on a previously unimagined scale.

Improvements in, and radical changes to, the way the commercial and social environment communicates with the user must take place in order for the global culture-complex to function. Design and designers will be entrusted with the task of finding the solutions to these problems.

In an attempt to address these challenges several UK design practices have established information design teams composed of product, interior, and graphic designers. This approach may well produce clearer information in print design or elegant, readable, electronic signage, yet as a strategic approach to a fundamental problem it is overly simple.

Neither the physical fabric of the designed environment (which includes print, packaging and product information) nor the cognitive abilities of the user is capable of carrying the weight of information that social, commercial and cultural trends will load upon them. The sheer scale and complexity of the problem renders it insoluble within the normal context of providing design solutions to design problems. Information can no longer be an interface between the citizen and the environment because the interface will become so complex and obstructive that the link will counter its own purpose.

The answer is not better information design but design for less information. What is required is less overt information. This is where those design companies cobbling together 'bolt-on' information design departments are so wide of the mark.

In order to address the information design problems posed by the global market a radical change in perceptions of what design is and how it functions must take place. 'The design' must cease to be the structure on which information is hung, or the medium through which it is transmitted, but must become in itself, the information.

UK designers frequently and sometimes without justification claim that design is about communication. This conveniently establishes design in the vanguard of communication industry services which are essential to the proper prosecution of commercial marketing strategy and therefore a compulsory purchase for client company marketing directors. In fact most of us perceive design as essential and central to the business of communication. However, there is a critical gap between our well-meaning words and the practice of commercial design.

As Saul Wurman (1991), designer of the American ACCESS information design system, said 'Despite the critical role that graphic designers play in the delivery of information, most of the curriculum in design schools is concerned with teaching students how to make things look good. This is later reinforced by the profession, which bestows awards primarily for appearance rather than for understandability or even accuracy. There aren't any (awards) for making graphics comprehensible'.

Design, to be effective, must enable customers to take actions based on informed decisions. Other approaches may produce wonderfully attractive designs, but must inevitably be contrary to the real needs of the global market and a waste of the potential that design and designers have to affect that market.

Solving the Tower of Babel syndrome and thereby making the global market intelligible to consumers represents the greatest challenge, and the greatest opportunity for designers in the next 20 years.

Changing the mind-set

The most difficult problem facing all European businesses in the globalizing market place is that of psychological adjustment. Managers and employees in every business and at all levels are facing considerable difficulties in defining their new roles in what has become a very complex commercial environment. Before designers and design companies can capitalize on the opportunities presented by the globalizing marketplace they must:

- adopt a global cultural mind-set
- adopt a marketing ethos, both in the manner in which they run their own business, and in the offer they make to their clients
- practise strategic marketing in order to cope with the volatility of global markets
- become strategically agile and flexible.

In adopting the global cultural mind-set, the geographical location of client base and supplier is unimportant. Changes in corporate culture; individual aspirations, prejudices and assumptions; business strategy and corporate positioning; are all of paramount importance. Practicalities like fluency in foreign languages are of course important, but an enthusiasm for foreign cultures, and a hunger for international exchange, exploration and travel all carry a much greater significance.

In the global market, attitude as much as aptitude, will become a significant criterion in determining fitness to serve.

The need for a marketing-led ethos

In the years since the end of World War II, the developed nations have created the most complex and powerful market-led society the world has ever seen.

No business can ignore the market, no service industry can avoid having a market-led strategy. By their very nature, design clients addressing global or wider European markets will be marketing-led and they will adopt ever-more sophisticated marketing tactics and methods.

These clients require proof that design in general, and a given solution in particular, can contribute to their commercial advantage.

Any service industry which hopes to supply creative marketing services to commercially sophisticated clients, must be at least as sophisticated as those clients in its own business ethos and practice. Anything less is a recipe for disaster.

Some designers have, in the past, mistakenly supposed that a marketing-led corporate ethos could not exist side by side with creativity. It is, however, both possible and desirable to retain creativity and to make those creative skills fit for the market place.

The volatility of the marketplace

In the past, markets have, for the most part, been well-defined, easy to identify and stable. It was the function of marketing and marketeers to try to predict changes and make commercial advantage out of getting the predictions right.

Factors such as the economic unification of Europe; the market hunger of the Far East; global cultural and aspirational convergence; the 'freeing' of Eastern Europe; the opening up of China; and the lifestyle aspirations of the huge and hugely wealthy Indian middle-classes, are putting vast amounts of energy into market dynamics.

The result of this energy release is a volatile eruptive and disruptive marketplace, in which markets constantly appear, bifurcate, fragment, reconstitute, and meld into one another, only to fragment once more. In order to cope with this volatility and the rapidly escalating growth in the rate of change, the old marketing function of trying to second guess the future and plan accordingly will no longer work.

Marketing must cease to be a separate activity which is added on to company function, and become an integral part of the corporate structure, ethos and activity. Only those companies adopting this strategic marketing ethos will have the strength, energy, agility, flexibility and speed of reaction necessary for survival and prosperity.

In the increasingly volatile market, producing a hard-and-fast plan and then blindly implementing it will be too risky for comfort. Establishing and implementing strategy should be an on-going process. The assumptions upon which strategy is based must be the subject of frequent and rigorous testing.

All strategic marketing needs strategic management. Management structures, decision-taking systems, implementation mechanisms, must be flexible, agile and energetic if companies are to survive.

The European market

Europe is now in the throes of the most energetic social, political and economic upheaval that it has experienced since the late 1940s. For this reason alone the wider European market deserves to be considered a special case within the global marketplace. Europe is also the most physically accessible of the global super-blocks for UK designers. It can be argued, however, that because of the rich cultural mix and residue of historical antagonism it is also one of the most difficult markets for UK design companies to address.

It is a common mistake to confuse the European market with the European Community (EC). While the countries which make up the EC are a significant part of the wider European market they are, in fact, only part of the story.

The European Community

At the time of writing the EC has twelve member states: The United Kingdom, Italy, France, Germany, Spain, Portugal, The Netherlands, Belgium, Luxembourg, Greece, The Republic of Ireland, and Denmark. Together these twelve states have a population of over 350 million people. In 1990, in terms of cross-border trading, the EC was a larger market than the USA and Japan combined. Furthermore, the total revenue value of cross-border trading within the EC more than doubled in 1990, from $14.83 billion to $32.2 billion. This rapid growth will continue throughout the next decade. Austria, Cyprus, Malta and Turkey have all applied for membership, and Switzerland is expected to apply in the near future.

The European Free Trade Area

Austria, Switzerland, Sweden, Finland, Norway, Iceland and Lichtenstein, are part of The European Free Trade Area (EFTA). The European Community and EFTA have already agreed in principle to facilitate the application of the free trade functions of the Community being applied across EC/EFTA borders.

In the medium to long term it is probable that many of the former iron curtain countries will either become full members of the community or form their own trading block which will embrace free trade with the EC/EFTA axis.

Whilst it is not possible to predict with any certainty how the whole picture will hang together, it is probable that by the year 2020 the wider European market will be home to between 470-550 million consumers.

Europe is the home market

For UK design companies, it is a mistake to perceive Europe as an export market; the wider European market is, or should be, the home market. The aim for all design companies should be to position themselves as European, based in the UK.

In the coming decades, there will, of course, be opportunities within the local market (UK), just as in the past there have been opportunities for design companies based in the UK provinces within their local markets. However, the really exciting opportunities will be available for those companies with the skills and resources to both move into the wider European market and begin to address the global markets.

The removal of trade barriers within the EC, and the probable developments in other European areas present an unprecedented range of opportunities to European commerce and industry. At the same time free trade within the community and the energy of the emerging market represents a danger for those companies which are slow to react. The message that the market is sending to design clients is simple: 'compete in Europe or loose your market.' Consequently during the 1990s, more and more client companies will be forced into proactive efforts to address the Euromarkets.

Most design clients will retain a large degree of local managerial autonomy, and local management will have the authority and budgets to commission design. At the same time the strategic commissioning criteria will, to a large degree, be dictated by Europe-wide strategic marketing considerations, and local decision-takers will report to Euro-chiefs. For many larger client companies the open market of wider Europe is already a fact, and their strategy and commercial tactics reflect this.

By definition Euro-active businesses are more proactive, commercially aggressive, and market smart than those 'in England, now abed'; they are also more sophisticated about the significance, function and process of design. Euro-active businesses put a greater value on designers and the design function than all but a few commercially aware UK businesses can or do. Their design decisions proceed from greater understanding; they have the corporate will and, significantly, the budgets to purchase design. Euro-active businesses represent the most attractive client base currently available to UK design companies.

There is already a discernible trend in Europe towards bigger commercial groupings in such areas as fmcg, retailing and the automotive industry. The bigger groups are absorbing and squeezing out the smaller. Legislation from Brussels will ensure and increase competitiveness among EC-based companies. Currently acceptable national monopolies are, or will become, commercially untenable and illegal.

These trends, together with the rationalization of some aspects of the design buying process onto a Europe-wide basis, will effectively reduce the number of clients available to design companies. Influencing the decision-makers will become much more difficult, and success will carry with it concomitantly greater rewards. The scale of jobs will increase, budgets will be larger and more prestige will accrue to the successful bidder. Only those design companies with the breadth of offer and depth of experience to undertake multi-national projects within the terms of reference set by the client's market, will be able to address the design projects set by the marketplace.

Conclusion

There can be no doubt that the cultural divisions delineating the nation state will remain. These divisions may, in fact, become more focused. Understanding and acknowledging these cultural, semiotic and iconographic differences within the context of multicultural markets will be essential for all businesses that wish to compete and succeed in the global marketplace. In the next two decades businesses of all types and sizes will turn to designers and design companies to help them understand and transmute that understanding into commercial advantage.

If design is to take its rightful place as the pivotal mechanism in global commercial communications, designers and design companies must make changes. These changes must occur in both the practice of the design business and in attitudes to the nature and function of design. In making these changes, fortune will favour the brave, and the over cautious will be relegated to footnotes in the history books.

Further reading

Hamel, Gary and Prahalad, C K (1991) Corporate imagination and expeditionary marketing. *Harvard Business Review*. July-August.

Hochstrasser, Beat and Griffiths, Catherine (1990) *Regaining Control of IT Investments*. London: Kobler Unit, Imperial College.

Mazur, Laura (1991) *Marketing 2000: Critical Challenges for Corporate Survival*. London: Economist Intelligence Unit.

Tibbetts, Joe (1991) *Developing Business in Europe*. London: International Design Marketing.

Wurman, Saul (1991) *Information Anxiety*. London: Pan.

Gary Powell/Sharp Practice

'Wisdom is the principal thing;
therefore get wisdom: and with all
thy getting get understanding.'

Proverbs

17 INFORMATION, RESEARCH AND INNOVATION

Liz Lydiate

This chapter is about the role of information and research in design consultancy, as part of the overall pursuit of innovation and creative excellence. Because it is now recognized that these qualities have no 'stand-alone' merit if they are not linked to the provision of an appropriate and effective solution which meets, or surpasses, the requirements of the brief, the collection and handling of information and intelligence have assumed a new significance.

The role played by commercial market research in relation to design practice embodies one of the greatest single changes in the nature of design consultancy during the last 20 years. Design consultancy is now able to offer clients highly sophisticated solutions to precisely defined marketing objectives, achieved through the synthesis of information and innovation. The quality of the result is directly influenced by the relevance, accuracy and breadth of the information which is collected and applied during the development of the design solution. In design consultancy, the collection of information and/or research material is required for a number of different reasons.

In relation to the consultancy's own marketing initiatives:

- researching/evaluating new markets, new services, new territories
- researching specific client prospects.

In relation to client companies:

- structure, personnel, objectives
- strengths, weaknesses, opportunities, threats (SWOT)
- field of operation
- competitors
- target markets.

In relation to individual projects:

- target market
- creative
- factual
- technical
- legislative
- product life cycle/planning
- ethical.

Only a small number of companies have dedicated research and information units, and it follows that, as with marketing, responsibility for information and research activity will be shared by many members of the organization. There is a huge, and constantly increasing, quantity of information and data available, and the acquisition of basic skills in research and information gathering and evaluation will pay great dividends. The two primary advantages are:

- relevance of the design solution to the objective addressed
- a broad sweep of new material to stimulate and inform the creative process.

Market research is of such significance that almost ten per cent of overall consumer expenditure is channelled into it by manufacturers and service organizations.

Most suppliers have direct access to their market, and collect much valuable information in the course of normal work. Design consultancies are in a different position, in that they serve in effect two markets: the client, and the client's market. Success of a design project is judged by its effect on the client's market, so the design consultancy must satisfy two audiences, and one of these at arm's length.

The practice of market research is defined by the European Society for Opinion and Marketing Research (part of the International Chamber of Commerce) as follows:

> *The systematic collection and objective recording, classification, analysis and presentation of data concerning the behaviour, needs, attitudes, opinions, motivations etc. of individuals and organizations (commercial enterprises, public bodies etc.) within the context of their economic, political and everyday activities.*

This definition is helpful, because it sub-divides the information-gathering activity into several distinct steps, breaks down the types of information which might be studied, and lists the principal contexts in which activity takes place.

Research principles

Much of this chapter is devoted to introducing a range of research techniques derived from formal academic and marketing practice. However, it is important not to lose sight of the value of effective research and information gathering in relation to creative work, and to emphasize the value of designers developing these types of skills.

Don't skimp on research

Time, access to a range of information sources (see below) and a fundamental acceptance of the role of research in all creative projects will pay enormous dividends in terms of quality and depth in the eventual solution.

Avoid settling too soon

When time is tight, it is tempting to go with the first acceptable solution. The first solution is not necessarily either the only one, or the best available, and it is desirable to search deliberately beyond the 'settling' barrier. The use of a quota system, or several different people (or teams) is helpful.

Create judgement yardsticks

The selection of personal, or company, 'icons' of excellence is an effective test for evaluating new solutions, and providing an impetus for further research.

Store and re-use information

Because information gathering is both time-consuming and specific, a lot of material is wasted because it is not gathered in a systematic way and stored for re-use. Within a company, information should be recorded and stored in a way which makes it available to all staff, and treated as a valuable resource.

Work around 'holes'

Sometimes work is held up by lack of information, and this can be avoided or mitigated by adopting the technique of working around a skeleton structure. This means other work can proceed around and beyond a missing piece of information, which can be slotted in later, as a module.

Develop information pragmatism

There is no merit in taking a long route through research if there is a simpler solution. Time can be saved through lateral approaches, and constructive use of cheek.

Learn 'multiple-choice' problem solving

The generation and consideration of all possible alternatives are powerful devices in addressing problem solving.

Within a consultancy, much competitive advantage can be gained if employees learn the discipline of analysing problems or choices and carrying out basic research on a range of available solutions before presenting the issue to his or her superior for a decision and, subsequently, action.

Types of research

Most commercially based research takes place in two areas of activity; their names belie the order in which they actually take place.

Secondary, or desk, research

This is first-stage research and is based on using existing sources; it can yield vast amounts of relevant and valuable information at low cost. Desk research can usefully include material collected internally, from the consultancy's own staff, from the client, and from libraries and other information sources.

Primary, or field, research

This usually develops and extends information collected at the secondary research stage, through a programme of activities which directly addresses the market under review. Primary research puts the supplier in direct contact with the consumer; in the case of design consultancy, this can be either the client or the client's clients.

Research planning

An extremely valuable overall principle for research of any kind (from evaluating a potential new market to finding a lost door key) is to remember 'if you don't know where something is, you also don't know where it's not'. This sounds so simple as to be unimportant, but it encompasses a sound principle, which is that research must be thorough and systematic, and must not stunt its effect by subscribing to false premises. It is also important to remember that negative information can be valuable, and useful conclusions can be reached through a process of elimination.

A basic plan for addressing and using research could be as follows; the example relates to the imaginary consultancy, Designco:

- assemble, review and assess existing relevant information, eg collect all available data relating to the fact that Designco's sales are falling and action is needed
- establish the proposal to be investigated, eg should Designco change its offer to meet suspected ongoing changes in the client base, and if so, how?
- acquire necessary additional data, eg undertake primary and secondary research on relevance of proposed new Designco offer to target client groups
- collate, and analyse data in relevant ways, eg what are the likely achievable commercial results of adopting the new strategy?
- make decisions on the basis of information gathered, eg Designco will set up a new division to provide specialized services to the professions

Checklist of topics for research on markets

- analysis of market potential for existing service(s)
- estimate of demand for new service(s)
- economic and business forecasting
- identification of market characteristics; internal and external influences - existing and potential
- analysis and prediction of trends in the market under review
- size of market (economic and demographic)
- location of market
- composition of market
- competition within market (market members competing with each other)
- competition for share of market as clients (eg other design consultancies addressing or likely to address a particular market)
- international implications
- legislative/procedural/code of practice restrictions
- ethical considerations
- consumer perceptions/level of satisfaction

- formulate and adopt an implementa-mentation plan, eg 40 per cent of staff to be redeployed to new division, working to strategic plan adopted by board, headed by a senior partner.
- evaluate performance in the light of objectives and projections, eg sales growth of x per cent to be achieved in eight months' trading in new area.

Briefing for research

Research for design must start from a clearly established brief. If the objectives are defined and agreed at the outset, everyone involved will be working towards the same goal:

- establish the extent and nature of the information to be collected; avoid collecting irrelevant information, or information for its own sake
- define the areas (geographic, economic or demographic) to be investigated
- state which variables are to be measured
- decide upon the required degree of accuracy in the results
- ensure that the brief does not prejudice the selection of research techniques and procedures ('don't know where it's not').

Scope of research

Within the parameters outlined above, research undertaken by or on behalf of design consultancies is likely to fall into one of two fields: research is for the consultancy's own business development, or work is undertaken as part of a client project.

Sources of information

It is easy to overlook or underestimate the usefulness of such an obvious source of reference as a library. Libraries have a great deal to offer, but effort may be required in the early stages in researching suitable libraries and building up user links.

Libraries

- General public libraries may not hold stocks of detailed business publications, but some larger ones (such as Birmingham Central Library) do. Public libraries can still be very useful for general research.
- Specialized libraries: eg City Business Library; University of Warwick Library (financial statistics); Statistics and Market Intelligence Library (includes Exporters Reference Library); National Art Library (V&A Museum); British Library.
- Colleges and universities; it may be possible to gain access to a nearby institution (particularly if you are an ex-student or visiting lecturer). Many educational establishments are considering providing research services on a commercial basis.

- Membership organizations, eg British Institute of Management; industry research associations, trade associations; professional bodies; Chambers of Commerce.
- 'In-house' libraries; consultancies can save time and improve performance with up-to-date reference books.

Newspapers and magazines

It is woth researching the specialist press for an overview on current trends/issues in the sector being investigated. Media directories, such as *PIMS*, *2-10* and *Editors*, list publications by industry sector, type of publication and by geographical area, and give addresses and contact information.

Other people

Never underestimate people's willingness to talk, particularly about themselves. Much good information can be obtained by telephone, by asking direct questions or simply by taking someone out to lunch with the express and pre-disclosed intention of picking their brains.

Checklist of topics for research on products/brands

- analysis of market for existing product/brand
- compilation of data on consumer perceptions/attitudes towards product/brand
- comparative studies of competitor products/brands
- brand equity research
- consistency of message analysis across brand activity (eg packaging, personality, name, advertising)
- product life-cycle/planning
- product line research/brand extension
- sales forecasting
- international implications
- legislative/trade restrictions
- ethical considerations
- consumer satisfaction
- finding new uses for existing product or production capacity
- assessing consumer reaction to proposed new product/brand
- market-testing proposed new product/brand
- contribution of design to marketing objectives

UK government agencies

- Information is available free of charge from government agencies such as the Small Firms Information Service, local TECs, the Central Statistical Office and the British Overseas Trade Board (which offers an Export Intelligence Service). This information is of its nature fluid and changing, but it is possible to find current resources through reference entry points such as the *International Directory of Published Market Research* sponsored by the BOTB, and *Government Statistics: A Brief Guide to Sources* (from the Central Statistical Office).
- The Design Council, Chartered Society of Designers and the DBA are all aware of published government material of direct relevance to the design industry.
- Overall reference books, such as *Sources of UK Marketing Information* give access to the full range of information available.
- In overseas national markets, equivalent information can be obtained; in the UK, the embassy or trade delegation is the best starting point.

International government agencies

Organizations as diverse as the United Nations and the International Monetary Fund publish research data; a selection of these include:

- Organisation of Economic Co-operation and Development (OECD)
  - Economic Surveys of Member Countries
  - Biennial Economic Outlook Review
- Statistical Office of the European Community
  - Economic survey of Europe
  - General Statistical Bulletin

- International Monetary Fund
  - Balance of Payments Yearbook
  - Direction of Trade (annual).

Reference publications

A visit to the City Business Library, or even the business reference section of a good bookshop, will reveal information which goes far beyond the well-trodden territory of the Times Top 1000; for example, Dun and Bradstreet's *Guide to Key British Enterprises* lists 25,000 companies.

Special reports

Companies frequently commission and publish research as part of their own marketing initiative. Perhaps the best-known example of this in the design field is the economic survey of design consultancies published by James Capel in 1987. The availability of special reports is usually promoted through the relevant trade press as part of the PR initiative.

Organizations

Trade associations are listed in the *Directory of British Associations* published by CBD Research. Organizations such as the Advertising Association, Confederation of British Industry, the Market Research Society and the Institute of Marketing (which publishes *Marketfact*, a weekly round up of news and information) can all provide a way in to pools of specialized data and information. Many such organizations provide a dedicated information service.

Do not overlook the basic but vital information which can be obtained from Companies House, and other useful resources such as the Trademarks Registry, where it is possible to obtain information required to avoid inadvertent duplication of concepts and symbols in corporate identity work.

Subscription-based market intelligence services

In addition to the generally available electronic information services, such as Ceefax, Oracle and Prestel, it is now possible for consultancies (and clients) to subscribe to IT-based market intelligence services, such as Harvest, Mintel, FT data service and Textline, which provide access to data through in-house computer terminals. Market intelligence companies also offer tailor-made research packages to meet more advanced requirements, and compile, publish and make available to subscribers data on special subjects on an ongoing basis.

Field research

Broadly speaking, field research moves from the general to the specific. It is a process which develops upon the material gathered during the desk research phase, and addresses the market or issues which are being studied in a precise and direct manner, in pursuit of particular results.

Field research divides into two principal types:

- quantitative research, which concentrates on numbers, volume, geographic and demographic breakdown of data
- qualitative research, which looks at impressions, perceptions, influences and reactions.

Design consultancies may carry out field research at various stages of a project; for example, in investigating consumer attitudes to an existing brand, prior to design development work, or in testing design concepts with a view to making this information available to the client as part of the design process. Information will also come back from clients as a result of other field research, and it is helpful to understand the principal techniques which may be employed or referred to.

Sampling

Research cannot address the whole population relevant to any particular issue; it approaches a sample, a carefully constructed representative section of the target group. The skill with which the sample is constructed directly influences the quality of the results. Samples are currently constructed using four principal techniques:

- ***systematic sampling:*** applying a set numerical formula to defining a sample, eg every tenth address from the electoral roll
- ***random sampling:*** every member of the group to be studied must have an equal chance of being selected, eg numbers drawn out of a hat
- ***stratified sampling:*** where the sample is composed of a number of smaller samples, which together reflect the composition of the group being studied, eg a company which employs 65 per cent men, 35 per cent women might construct a representative employee sample by researching an equivalent male/female ratio
- ***cluster sampling:*** approaches respondents in relation to a particular topic or location, selecting a sample from within this pre-determined group, eg shoppers using selected branches of a chain of stores.

The method by which the sample is constructed is known as the sampling frame. Once the sample has been constructed, various different methods may be used to obtain a range of data.

Pilot research

A pilot study is a limited, preliminary piece of research used to test the feasibility of a larger-scale undertaking. In order to be useful, the pilot study must follow as closely as possible the objectives, parameters and methodology identified for the proposed follow-on project. The findings of the pilot are used to inform, refine and develop the larger undertaking, in order to maximize potential results.

Questionnaires

Questionnaires are used to collect specific information in a controlled, standard manner from pre-determined groups of people (the sample). Questionnaires can only be expected to produce a limited response, and the design of the questionnaire and the size of the population (ie group(s) to whom it is addressed) must be determined to take account of this and still yield sufficient answers to produce a workable result. The Market Research Society lays down useful guidelines for the minimum acceptable contents for a research report based on a questionnaire, and the following may be used as a checklist for undertaking this type of study:

- overall purpose of the research
- specific objectives of the questionnaire
- details of the principals (commissioner) and the researchers
- population to be addressed
- size of sample and method of construction
- methodology
- timing
- the questionnaire itself
- location of interview
- database
- findings.

Interviewing and vox pop
Interviews may be carried out by telephone, by appointment (in groups or one-to-one), knocking on doors or stopping people in the street. The research must follow a pre-planned route in search of specific information, so that all findings can be linked up and systematically analysed.

Vox pop involves interviewing people, often selected at random, and recording their answers to planned questions. It is a useful method of conveying people's attitudes and reactions, and is often used to overcome intransigence and pre-judgement of issues influencing the marketing strategy.

Experimentation and observation
This can involve the test marketing of new products or the study of consumer reactions in controlled situations.

Experimentation is complex and expensive to set up, but if used more widely, may make an important contribution to overcoming consumer frustration at manufacturers' frequent neglect of design considerations. Similarly, observation is also directly relevant to the design process, particularly in store design, circulation systems, crowd management, product development and brand packaging.

Syndicated research
Syndicated research is carried out where there is an identifiable need for information which can be used by a number of companies and organizations operating in a particular field, and the provision of information becomes commercially viable. Material is gathered by providers of syndicated research in four principal fields:

- trade audits
- television audience ratings
- consumer panels
- omnibus surveys (eg Gallup Poll).

Qualitative research technqiues
Qualitative research techniques vary widely, because the objective is the collection of conceptual and attitudinal information; this type of research must be carried out by skilled staff if it is to yield reliable results.

The techniques of qualitative research are frequently used within design consultancies as an aid to creative thinking and research, and may include activity derived from the following starting points:

- group discussion (including brain storming)
- one-to-one interviewing; probing behaviour, attitude, opinion and needs through non-directive discussion
- use of stimuli and 'starting point' techniques, eg ink-blot tests, picture interpretation, character definition in terms of cars, animals etc.

The role of research in evaluating design

All of the research techniques described can play an additional key role, in evaluation, or performance assessment.

Evaluation has a very important contribution to make in demonstrating and quantifying the effectiveness of design. If clear targets and objectives have been established at the outset of each project, as described earlier in this book, it is both possible and advisable to try to establish exactly what has been achieved, and to what extent, once the project has been completed.

Design performance evaluation can best be done with the total co-operation of the client, who of course stands to gain valuable information from the exercise. However, because this is a relatively new concept, it needs to be presented to the client as a suggestion at an early stage, and it will take consistent and concerted effort by designers and design consultancies to establish evaluation as a norm.

Further reading

De Bono, Edward (1990) *Lateral Thinking*. London: Penguin.
De Bono, Edward (1990) *Lateral Thinking for Management*. London: Penguin.
Evans, Roger and Russell, Peter (1990) *The Creative Manager*. London: Unwin.
Foster, Timothy R V (1991) *101 Ways to Generate Great Ideas*. London: Kogan Page.

Simon Stern/Sharp Practice

'There was hardly a more pervasive theme in the excellent companies than respect for the individual. What makes it live is the plethora of structural devices, systems, styles and values, all reinforcing one another so that the companies are truly unusual in their ability to achieve extraordinary results through ordinary people.'

Peters and Waterman
In Search of Excellence

18 HUMAN RESOURCES
Sue Young

Human resource management as a recognized function has developed out of a realization that people are the most important resource in a successful business. In a service industry such as design, the human resource factor is particularly important; it could be argued that a design consultancy's only resource is its people. In a business where creativity, vision and service are the products bought by the client, ability to deliver on the offer derives from the qualities and skills of individuals.

The human resources management (HRM) function

It is important to separate the function of HRM from the management role, that is the person in the organization who has overall responsibility for managing human resources. This is particularly so in the design industry where very few companies can afford to have a full-time human resources manager, and it is common for different aspects of the function to be divided between a number of individuals.

In recognition of the strategic contribution which people make to a company, there has been a trend in recent years away from the description 'personnel' to 'human resources'. The term 'personnel' has become downgraded and tends to be associated with the more bureaucratic and administrative aspects of the HRM function, and seen as a 'dead' cost to the business. In contrast, the HRM philosophy sees people not as a cost but as a resource and opportunity. It sees the HRM function as the integration of the organization and its goals with the deployment of human resources to meet those goals. Philosophically this chapter takes very much the HRM approach.

The Institute of Personnel Management (IPM) defines the personnel management or HRM function as 'A managerial function, personnel management seeks to develop into effective organizations those men and women who make up an enterprise, enabling each to make their best contribution to its success'.

As people are design's most important asset, there is much to be gained in the design industry from learning and using good people management techniques. The HRM function would typically include the following activities:

- organization design and development
- recruitment and selection
- induction
- training and development
- employee communications
- pay and benefits determination and negotiation
- payment administration

- personnel information and records
- disciplinary procedures, grievances and disputes
- health, safety and welfare.

This chapter concentrates on the more managerial aspects of HRM, with a brief overview of the main legal and contractual issues of employment.

The job description

At a day-to-day operational level, organizational design and development is implemented through preparation of individual job descriptions, which set out the specific tasks, duties and relationships of a post.

No job in an organization is completely self-contained. Each job has working relationships with others; upwards, downwards and sideways to parallel jobs. All these links help to shape the individual job, and all the individual jobs together form a structure, the shape and style of which have a strong influence on the character or culture of the organization as a whole, and on its effectiveness.

Job descriptions are beneficial to any organization in:

- formally defining individual responsibilities
- providing a framework against which organization structures can be reviewed and developed
- creating a basis on which roles can be evaluated
- acting as a foundation for performance appraisal
- recording information on which to base a person specification.

Preparing the job description

A job description is a description and a definition of a particular role within an organization. There is no single format for a job description (see example on page 185). Job descriptions represent a dynamic, evolving and essential tool for management at every level of responsibility, including the board. All job descriptions should be reviewed on a regular basis and updated where necessary.

Recruitment and selection

The first question to be asked before embarking on the recruitment process is 'Does the role exist and what is the real nature of the job to be done?'. Too many design consultancies make the mistake of not thinking about the issue of the role clearly enough. As a result they can be easily seduced by the 'chemistry' of a particular personality and recruit a person without really thinking about the job requirements and attributes needed. It is essential to understand the nature of the role before adapting it to individuals, and not the other way round.

The job description is the manifestation of a clearly defined role. A detailed and attractive job description has the added benefit of being a persuasive selling aid in the recruitment process. It demonstrates a thorough and professional approach, indicative of a company that really values its people and their contribution.

Preparing the person specification

The 'person specification' is arguably the most important single document in the entire selection process, and must be linked closely to the job specification. It is a detailed definition of the abilities, attributes, experience, skills and knowledge required to do the job, and creates a picture of an ideal candidate. Ideal candidates are rarely found, but there are two important reasons for this approach:

- it provides a standard against which a range of candidates can be assessed
- it assists in considering training/development plans to close the gap between what the organization needs from the candidate and his/her current skills and experience.

The following factors may be considered in drawing up a person specification:

- previous relevant experience
- other attainments/ qualifications
- skills
- work interests
- work attitudes
- personality
- circumstances (if relevant to the job, eg ability and willingness to travel)
- physical make-up (if relevant to the job requirements).

Each factor can then be graded; essential, desirable or negative.

Essential attributes are qualities, skills, qualifications or experience, without which the job could not be performed successfully. Even if candidates perform well in other respects they should not be considered for an appointment if they are below standard in any essential attribute.

Desirable attributes are qualities or experience which the selector would like, or prefer a candidate to possess. A candidate who is below standard on one of the desirable attributes, but is suitable in other respects, may prove to be successful in future performance on the job.

Negative attributes are factors which constitute a handicap to effective performance.

The person specification can be used to:

- prepare a vacancy advertisement or to brief recruitment consultants
- prepare interview assessment criteria and plan questions
- formulate training and development programmes for individuals to help them perform their job more effectively.

Elements of the job description

1 Job title

2 Reporting relationships
- subordinates
- colleagues
- superiors

3 Organization structure
- department
- division
- company

4 Main purpose of role
- objectives

5 Key tasks
- short term
- long term
- timetables

6 Management style
- company style
- immediate manager

7 Available resources
- people
- money/equipment

8 Chief responsibilities
- staff
- money/equipment

9 Constraints
- financial limits
- controls and checks
- authorizations

10 Working conditions
- hours of work
- travel requirements
- physical demands

11 Prospects
- promotion opportunities
- training
- job satisfaction/status

12 Rewards
- pay/bonus
- profit participation
- pension
- fringe benefits

13 Holiday entitlement

14 Notice entitlement

Different methods of recruitment

There is no golden rule about the best method of recruitment. Whatever the method, the aim should be to create a pool of candidates with most, if not all, of them suitably qualified on paper for the job. The most popular methods of recruitment include:

- recruiting internally
- personal contacts, including existing employees
- colleges (in the case of junior designers)
- advertisements, usually in the trade press (but for more senior managerial appointments, could be national)
- recruitment agencies
- executive search consultants (headhunters).

Vacancy advertisements

There are a number of points to consider when recruiting through advertising:

- choice of media: consider what the candidates are likely to read
- the information in the advert counts for more than the way it is presented
- the advert should start with a clear headline of what the job is; don't use gimmicks
- give as full a job description as possible in the space allowed
- unless there are very good reasons for not giving salary and benefits, do so
- state essential and desirable requirements; avoid restricting it to an age band without very good reason
- watch out not to offend the Sex Discrimination or Race Relations Acts
- position and size of the advert are relatively unimportant to readers
- tell people clearly how to respond. In most cases this will be to send a CV with a covering letter
- always handle enquiries and reply to applications quickly. Never fail to reply as it gives an unfavourable impression of the company.

Briefing recruitment consultants

The brief to a recruitment consultant should include:

- background about the company, client profile, development in recent years, culture and style of working
- a description of how the company is organized (organizational context for the job)
- a detailed job description in writing
- the person specification in writing.

The recruitment process

Whatever your chosen method of attracting applicants, the recruitment procedure will go through a number of stages.

1 Shortlist

The shortlist will contain applicants who should, at a minimum, meet the essential criteria. Be sure to reply promptly to all applications.

2 First interviews

Try and arrange a set period over which the interviews will be held, say one or two weeks. Condensing the initial interview programme enables easier comparison and, if necessary, refinement of the person specification.

The aims of the first interview should be:

- to find out from the applicant information which will lead to an initial assessment of whether they are capable of doing a good job
- to allow the applicant the opportunity to find out any information they want about the company.

3 Second and subsequent interviews

As a result of the first interviews there should be a shortlist of no more than two or three people who the employer believes could be capable of doing the job. The aims of further interviews can be several:

- to allow the applicant to meet others in the organization that they will be working with in order to get others' assessment of the individuals
- to probe more deeply into particular areas of experience or skills
- to begin the process of negotiation by finding out more about the position the candidate is in (what package they are now on, what their expectations are etc.).

If the recruitment process is properly planned there should be no need to go beyond three interviews before making a decision. Sometimes it may be necessary to go to a further meeting for discussion of unresolved elements of the package.

4 Evaluation and decision-making

The more senior the appointment the more critical it usually is for several people in the organization to share in the decision-making. It is important that all involved in the interviewing and assessment process are working to and agreed on the same criteria. Within that overall framework individuals may take responsibility for assessing issues falling within their particular area of expertise or concern.

5 Negotiation

Salary and package details will form part of the job specification and may be stated in the advertisement, or as part of a brief to a recruitment consultant. In some cases this may be a range, where there is considerable variation in the market rate.

One of the aims of the interviews should be to obtain a clear understanding of the candidate's position and expectations on the main aspects of a package. On aspects where there are clearly differing views, it is essential that both views are fully explored and shared to enable the employer to form a view on what offer would be acceptable. So, when a formal offer is made to the candidate there should be no surprises.

6 The offer

The employer has a statutory obligation to provide an employee with a written statement of full particulars of the main terms of employment within 13 weeks of starting the job, but, wherever possible, these should form part of the formal offer.

This written statement should cover the following as a legal minimum:

- title of the appointment
- date when the employment commenced
- whether time with a previous employer counts as part of this employment
- the period of the contract (if it is fixed)
- salary and intervals of payment
- pension arrangements, including whether the company has contracted out of the State scheme (SERPS)
- holiday and holiday pay entitlement
- sick pay entitlement
- any fringe benefits
- disciplinary and grievance procedure
- procedure for termination and notice periods.

The employer may want to incorporate additional terms such as the following:

- ***confidentiality:*** employees must observe strict confidentiality in relation to any knowledge concerning the employer's business, and do the same for their clients
- ***restrictive covenants and restraint of trade:*** contracts may restrict the ability of former employees and partners to compete in a similar business within a particular

geographical area or for a particular period. Legal advice should always be sought on such a restriction (by both sides) because, if any part of it is unreasonable, the whole of it will be unenforceable

- ***intellectual property rights:*** copyright usually belongs to the originator of work (see chapter 15), but copyright in work made as an employee usually belongs to the employer, and the position on copyright ownership should be spelled out in the employment contract
- ***reference to the job description or a list of duties and responsibilities:*** discussion of this should have formed part of the recruitment process, but written confirmation is an important foundation for an effective working relationship between consultancy and employee.

The employer should always seek professional legal advice on any contractual negotiations but these guidelines should help in thinking through the issues in advance and result in a more effective working relationship with a solicitor. It is often advisable for employees to take legal advice before signing a contract, as the true meaning of contractual terms is sometimes not clear to the lay reader.

Interviewing techniques

There are a number of techniques which can help improve the quality of information gained from interviewing.

Have a plan

It is best for the interviewer to have a structure for the interview, outline it to the candidate, and then follow it. Explaining the structure at the outset helps put the candidate at ease because they know what to expect, and enables the interviewer to control the meeting. (An example of an interview plan is set out on page 189.)

As part of the plan the interviewer should prepare questions in advance. Not only does this allow the interviewer to retain control but it enables comparison of candidates on how they answer the same questions.

Develop rapport

Be natural and friendly with the candidate. Make them feel relaxed so that their behaviour is natural. This is particularly important at the outset of the interview when the candidate is likely to be nervous.

Take notes

Notes help the interviewer to remember what was said. Keep them short to avoid lengthy breaks in eye contact and listening, and perhaps expand them immediately following the interview while the information is fresh.

Use open and closed questions

The purpose of the interview is to get the candidate talking and to allow the interviewer to find out more about him or her. Open questions are a way of doing this and are particularly useful at the early stages of the interview. A common feature of open questions is that they cannot be answered with a short answer like 'yes' or 'no'. For example:

- 'Tell me about your time with Smith & Jones'
- 'What do you know of our competitors?'
- 'How would you feel about working unsocial hours?'

Closed questions require only a few words to answer them. Use them sparingly to probe a particular point or to clear a specific issue. For example:

- 'How long were you with Smith & Jones?'
- 'Do you know how many direct competitors we have?'
- 'Are you happy to work the occasional weekend?'

Leading questions
Leading questions are generally best avoided; the candidate will almost certainly give the answer they know the interviewer is looking for.

Talking time
The purpose of the interview is to provide information on the candidate; so he or she should be doing 75 per cent of the talking and the interviewer only 25 per cent.

Prejudice and bias
Everyone has their own subjective biases and prejudices. The interviewer's judgement of the candidate should be based as much as possible on the facts, and not on their individual feelings and personal reactions. Equal opportunities legislation has many practical implications for the recruitment process and special care needs to be taken not to fall foul of the law in:

- deciding on the person specification (the criteria)
- drawing up the shortlist
- conducting the interview
- making job offers.

Relevance
It is important to ask questions relevant to the job on offer, rather than just ramble over issues which the interviewer finds personally interesting.

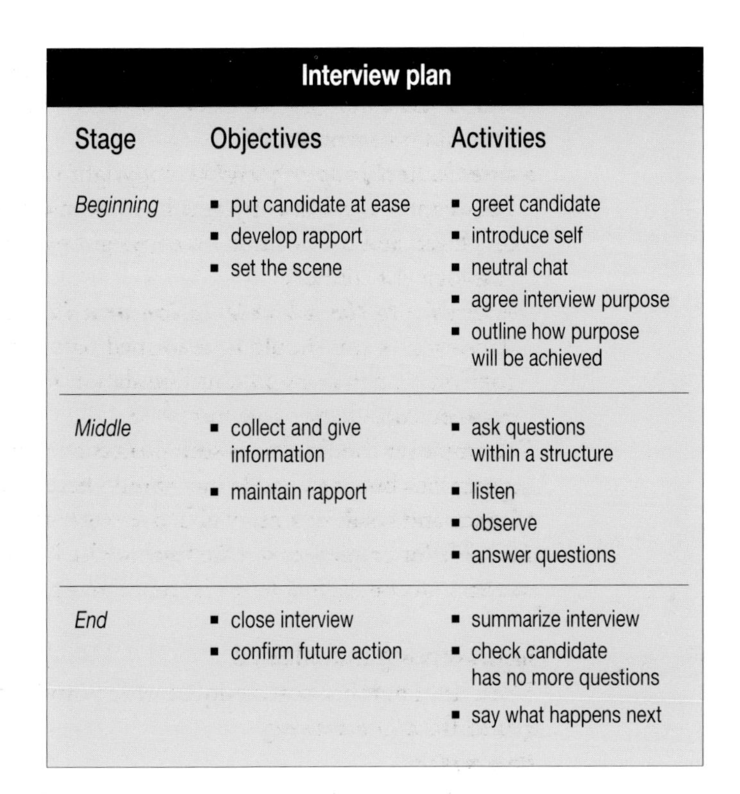

Interview plan

Stage	Objectives	Activities
Beginning	▪ put candidate at ease ▪ develop rapport ▪ set the scene	▪ greet candidate ▪ introduce self ▪ neutral chat ▪ agree interview purpose ▪ outline how purpose will be achieved
Middle	▪ collect and give information ▪ maintain rapport	▪ ask questions within a structure ▪ listen ▪ observe ▪ answer questions
End	▪ close interview ▪ confirm future action	▪ summarize interview ▪ check candidate has no more questions ▪ say what happens next

Induction

Starting a new job can be a stressful experience for any employee, and experience shows a much higher rate of resignations and dismissals during the first few weeks or months than at any other time. A common cause of failure during the first few days and weeks of employment is a sense of disillusionment. An optimistic view of the job is gained during recruitment and selection, but the reality often does not match expectations. Effective induction therefore starts as part of the recruitment process and aims to give a realistic and accurate impression of the company and the job role.

Pre-employment action
Actions at the recruitment stage for successful induction include:

- providing realistic information about earnings, bonuses or overtime
- showing the nature of the work and, if at all possible, taking the candidate around to see the workplace

- introducing the candidate to people with whom he or she will be working closely
- documenting in the formal offer a clear description of rights and duties
- clear instructions sent in advance to new employees about their first day at work; eg the time, place and person to whom they should report, car parking arrangements, details of any documentation they need to bring with them and so on.

Starting the job

The new employee's primary concern is with immediate, practical matters, such as finding their way about and meeting their new colleagues. It would be easy to overload them with a mass of information and detail about the company and the job.

The person to whom they will report must know of their arrival and be clear as to what to do with them. Reception and staff generally should be encouraged to be friendly and helpful. It is a good idea to assign someone of the same sex and age range to act as the 'starter's friend' in the first few days. The first day will also include an initial introduction to the work itself. There are several points to be borne in mind:

- the starter should not be overloaded with information; the full learning process, even for a simple job, should be spread over an extended timescale, with the emphasis on the first day on mastering a few basic tasks
- more senior level starters will need to spend time on familiarizing themselves with the organization and people: the company's culture, values and priorities
- all starters should be kept busy. There is little worse than having nothing to do and sitting conspicuously inactive at an empty desk.

The first weeks and months

Induction does not finish at the end of the first day. The whole process of becoming acclimatized to a new job takes weeks or months. The exact timescale varies from a minimum period of three months, where the work is simple, to considerably longer for more complex jobs.

It is crucial to the success of any induction that a friendly and supportive approach is adopted by the manager. A pro-active approach should be taken to identify any problems and they should be addressed as soon as possible.

If a probationary period is agreed, its existence and length should be made clear at the outset, and attention given to formally confirming the employment and giving feedback on progress as soon as the probationary period has elapsed.

Performance appraisal

Staff appraisal is an important element in successful development and motivation.

Done well, a staff appraisal can ensure the maintenance of good working relationships and motivated individuals.

Done poorly, a staff appraisal can, at worst, have a demotivating effect and, at best, be seen as a somewhat pointless management ritual.

The objectives of a performance appraisal process are to:

- review past performance against previously determined objectives
- discuss means of improvement and development of potential by setting specific and realistic targets and means of measuring their achievement, and identify how this can best be supported and encouraged
- ensure the maintenance of a good working relationship
- assess any need/opportunity for training or career development.

The focus should be forward looking, emphasizing development and improvement.

Salary review

Frequently there is a direct link between salary grading and performance appraisal, with salary increase following directly after appraisal. Such a direct link will not encourage a free and frank discussion between the appraisee and his or her manager. The best approach is to separate salary from performance appraisal by six months, thereby creating the opportunity for salary to reflect any change in performance since the appraisal.

This approach turns the link between appraisal and salary review from a shackle into an incentive.

Principles of effective appraisal

An effective appraisal process should:

- balance the review of past performance with planning for the future
- give the employee the opportunity for self-assessment: how they think they've done, areas that have presented problems and what they feel are their development needs. A simple form completed prior to the meeting can be a useful prompt.
- focus on the job and its requirements, rather than the person; the job description provides a useful starting point
- the formal appraisal should be an opportunity to stand back from the day-to-day and take stock; it is not a substitute for good management and on-going communication.

Appraisal should be an on-going process, and its frequency must be determined according to the needs of the job. A new employee, someone who is having problems, or a person in a developing, changing role may benefit from three-monthly meetings. Where somebody is established in their job, seems to be doing satisfactory work and there is little need for change, an annual appraisal may be sufficient.

Formal appraisal is an excellent opportunity to assess the status quo from both points of view and provides an opportunity to present staff with development opportunities and challenges which will help to keep them motivated and committed to the company.

Training and development

Most learning takes place informally at work without people necessarily thinking of it as training. People try things out; reflect on their experience; seek (or are given) advice; and get information, perhaps from observing others or doing their own research. However, training as a more formal intervention has the specific purpose of providing employees with the skills, knowledge and attitudes required to meet the company's objectives. It is a systematic process to ensure that learning by individuals meets the needs of the business.

Training and development are also powerful motivational factors. Design consultancies have much to gain from the stimulation, new information and drive which can result from good external or formal internal training.

Identifying training needs

A training need exists when there is a gap between the present skills and knowledge of employees, and the skills and knowledge they require, or will require. Training needs fall into a number of categories:

- individual training needs, identified through the performance appraisal process
- needs in the company overall, where, for example the development, and perhaps

survival of a business depends upon a much greater willingness on the part of managers and employees to accept change
- needs of specific groups of people; for example, the introduction of new technology in the studio or the need for designers to have improved presentation skills.

Choice of training methods

Choice of training methods will depend on a number of factors including the trainee themselves, the nature of the learning objectives and the organizational context in which training is to take place. Possible training structures fall into five main categories:
- training within the present job environment
- other planned in-company activity
- planned external activity (other than courses)
- internal courses
- external courses.

Disciplinary procedure and termination of employment

This section aims to give an overview, as an aid to understanding, of legal and contractual issues in relation to disciplinary procedure and termination of employment.

Contracts and other legal undertakings should never be set up without seeking professional legal advice.

Different types of termination

The law relating to termination of employment has had a profound effect on the development of disciplinary procedures. There are several different circumstances and methods through which employment may be terminated.

Termination with notice

Either party is entitled to terminate a contract of employment by giving notice, and the length of the notice should be set out in the contract of employment. The law requires that all employees be given a statutory minimum period of notice determined by how long they have been working for the company.

Termination without notice: summary dismissal

A summary dismissal occurs where the employer terminates a contract of employment without notice on the grounds that the employee is in breach of an express or implied term of the contract, ie guilty of gross misconduct.

A dismissal made without notice, or with inadequate notice, in circumstances where proper notice should have been given, is considered a wrongful dismissal. The expression also covers dismissals which are in breach of agreed procedures.

Unfair dismissal

After two years' continuous service every employee, with a few exceptions, has the right not to be unfairly dismissed. Certain types of dismissal will automatically be regarded in law as unfair:
- dismissal related to trade union membership or activities, or non-union membership
- dismissal on the grounds of pregnancy
- discrimination under the Sex Discrimination Act 1975 or the Race Relations Act 1976
- discrimination against a person who has a conviction which has become spent under the Rehabilitation of Offenders Act 1974.

Framework for disciplinary procedure

Offence	Penalty	Management Involvement
Minor misconduct	Verbal warning	Supervisor/ immediate manager
	Written warning	Department manager
Repeated minor misconduct or serious misconduct	Final written warning	Department manager/ personnel manager
	Transfer, demotion or suspension	Senior manager and personnel manager
Repeated minor misconduct or gross misconduct	Dismissal	Senior manager Personnel manager

Fair reasons for dismissal

The employer must be able to prove reasonable grounds for the dismissal under at least one of the following categories.

Competence

An employee may be dismissed if the employer has reasonable grounds to believe that he or she is incapable of carrying out his or her duties. However, tribunals will want to satisfy themselves that all reasonable measures possible were taken by the employer in order to give the employee the opportunity to improve.

Ill health

An employee can be dismissed if unable to carry out his or her duties under the terms of their contract of employment. People who are registered as disabled are entitled to special consideration.

Conduct

Where there is serious misconduct, eg theft, an employee can be dismissed without warning. In the event of employers dismissing an employee for other types of misconduct there is a need to prove the dismissal to be fair. In the case of suspected dishonesty the employer has to prove that he or she entertained a reasonable suspicion amounting to a belief in the guilt of the employee at the time.

Redundancy

It is not enough to show that it was reasonable to dismiss an employee; the employer has to convince a tribunal that he or she acted reasonably in treating redundancy as 'a sufficient reason for dismissing the employee'. If grounds are chosen other than 'last in, first out', employers must be able to prove that other criteria chosen are reasonable and that they have been applied rationally and objectively.

Disciplinary procedure

The law lays down a procedural framework indicating outline stages a disciplinary procedure should follow. The majority of problems can be resolved if dealt with firmly and fairly at an early stage. The manager should discuss with the employee the reasons for their lack of performance, and explain why it constitutes a problem for the company. A way forward should be initially agreed, with the employer providing any reasonable input or help required.

Redundancy

Employees are considered to have been made redundant if their dismissal is due to their job disappearing as a result of either internal re-organization, or reduction in level of business. It is not linked to lack of individual performance in the job.

Minimum statutory redundancy payments must be made, calculated as follows:

- one and a half weeks' pay for each year of employment in which the employee was aged between 41 and 64 (59 in the case of women)
- one week's pay for each year of employment in which the employee was aged between 22 and 40
- half a week's pay for each year of employment between the ages of 18 and 21.

Managerial considerations

Making redundancies is never pleasant and the aim should be to manage the process in a way which is considerate of the potentially traumatic implications for the individual. Managers should be open and honest and thus avoid sustained rumour-mongering. Internal communications become even more important than external communications; there can be little worse than employees reading about proposed redundancies in the press before being told by their employer.

Afterwards, as much help and support as possible should be given in helping individuals find alternative employment, such as providing office facilities for a limited period, giving references, or making counselling services available if required.

Redundancy practice throws the issue of people management in design into sharp relief. No other area of human resources, if handled poorly, has such dramatic potential to damage a company's long-term reputation in the industry.

Further reading

Farnham, David (1990) *Personnel in Context.* London: Institute of Personnel Management.

Fowler, A E (1983) *Getting Off To a Good Start.* London: Institute of Personnel Management.

Handy, Charles (1986) *Understanding Organizations.* London: Penguin.

Harper, Sally (1987) *Personnel Management Handbook.* Aldershot: Gower.

Herriot, P (1989) *Recruitment in the 90's.* London: Institute of Personnel Management.

Peters, Thomas J and Waterman, Robert H Jnr (1984) *In Search of Excellence.* London: Harper & Row.

Randall, G, Packard, P, and Slater, J (1984) *Staff Appraisal: A First Step to Effective Leadership.* London: Institute of Personnel Management.

Shackleton, V J (1989) *How to Pick People for Jobs.* London: Fontana.

Stewart,V and Stewart, A (1978) *Practical Performance Appraisal.* Aldershot: Gower.

Taylor, B and Lippitt, G L (eds) (1983) *Management Training and Development Handbook.* New York: McGraw-Hill.

Torrington, Derek and Hall, Laura (1991) *Personnel Management: A New Approach.* Hemel Hempstead: Prentice Hall.

Index

L

M

N

O

P